Textile Science and Clothing Technology

Series editor

Subramanian Senthilkannan Muthu, SGS Hong Kong Limited, Hong Kong,
Hong Kong

More information about this series at http://www.springer.com/series/13111

Subramanian Senthilkannan Muthu
Editor

Textiles and Clothing Sustainability

Sustainable Fashion and Consumption

 Springer

Editor
Subramanian Senthilkannan Muthu
Environmental Services Manager-Asia
SGS Hong Kong Limited
Hong Kong
Hong Kong

ISSN 2197-9863   ISSN 2197-9871   (electronic)
Textile Science and Clothing Technology
ISBN 978-981-10-9536-8   ISBN 978-981-10-2131-2   (eBook)
DOI 10.1007/978-981-10-2131-2

Printed on acid-free paper

This Springer imprint is published by Springer Nature
The registered company is Springer Science+Business Media Singapore Pte Ltd.

Contents

The Role of the Retailer in Socially Responsible Fashion Purchasing

Alana M. James and Bruce Montgomery

Abstract In recent years there have been a number of dramatic changes in the fashion retail arena, not only have consumer needs changed but also the way they purchase fashion products. The emergence of the fast fashion business model allowed consumers access to new collections every few weeks, with some retailers now moving to up to 12 collections per year. Whilst this approach reflects the very antithesis of *fashion;* constantly renewing itself and offering new trends every season, the pace of the supply chain has also had to increase to meet demand. This constant access to new fashion products has adapted the wants and needs of the consumer, with quantity rather that quality being preferred. However, it is this acceleration of the supply chain that has lead to a number of social and environmental issues occurring. Opposing this speed of consumption is socially responsible purchasing, where ethical considerations are taken into account during the design and production of fashion. Both consumers and retailers alike are beginning to adopt this approach evidenced in both corporate social responsibility strategies and ethical purchasing behaviour. However, there are key issues currently preventing a fundamental change in the industry such as a lack of incentives for responsible purchasing behaviour and poor aesthetics. It is these issues that need addressing in order to push the industry towards a more socially responsible future. The positioning of the responsibility for the future of the industry is a highly debated topic, with both consumers and retailers often being assumed as leaders. This chapter explores the role of the retailer in this debate and the powerful position they are in as the middleman between the purchasing consumers and the manufacturing suppliers. In order to influence consumer behaviour towards a more socially responsible approach, retailers need to engage their consumers in an innovative way. Their

A.M. James (✉)
School of Textiles and Design, Heriot-Watt University, Galashiels, UK
e-mail: a.james@hw.ac.uk

B. Montgomery
Faculty of Arts, Design and Social Sciences, Northumbria University,
Newcastle upon Tyne, UK
e-mail: Bruce.montgomery@northumbria.ac.uk

© Springer Science+Business Media Singapore 2017   1
S.S. Muthu (ed.), *Textiles and Clothing Sustainability,*
Textile Science and Clothing Technology, DOI 10.1007/978-981-10-2131-2_1

unique role is to inspire their consumers, using fashion as the communication vehicle for a drive towards future change.

Keywords Ethical fashion · Consumer purchasing behaviour · Fashion purchasing process

1   Background and Context

The fashion industry is one of the largest consumers of natural resources in the world. In order for the industry to operate, it requires large quantities of many of these resources such as water, for example, which is needed in order to grow and produce cotton based clothing. To produce a basic cotton t-shirt, for example, requires 2700 L of water throughout the manufacturing process (WWF 2015). Another heavily utilised natural resource in the fashion industry is oil, which is used to produce man-made fibres such as polyester. The draining of many natural resources and the consequential damage to the environment is in some cases irrevocable. Factory operation and transportation of goods causes toxic carbon emissions to be emitted, polluting the air and contributing to the long list of negative environmental consequences, which are as a result of the production of garments. In addition to the impact on the environment, there are also social issues to be considered.

Retail historically has tended to focus on the operational functions of primary importance to retail such as the brick and mortar stores, warehouses and distribution centres and stakeholder engagement. Latterly, however, retail is having to become more conscious of making the garment supply chain transparent and consider for the first time integrating its sustainability agenda throughout the business and on to the consumer.

Within the long and complex process of the fashion supply chain, dozens of factory workers, dyers and processors contribute to the production of fashion products. All of these human beings require a safe working environment and to be paid a fair living wage, however this is not always the case. In April 2013, the deadliest accidental structure failure in modern human history occurred (BBC 2013). Rana Plaza was a commercial building in the Savar region of Dhaka, Bangladesh and housed many clothing factories employing thousands of workers. Despite warnings regarding the safety of the eight-storey extension on the top of the building, 1130 workers were killed when the building collapsed. In the days leading up to the collapse, factory owners were advised to vacate the building due to cracks appearing the walls, however, due to pressure to complete orders, factory workers were told a months wages would be docked if they did not turn up for work the following day (Devnath and Srivastava 2013). Ignoring this advice caused the fatalities and injuries to a further 2515 people (Butler 2013). Within the factories in the Rana Plaza complex clothing for high-street brands such as Zara, Mango and Primark were being produced at the time of the collapse (Nelson and Bergman 2013). What all these brands have in common is that they are fast fashion retailers who aim to bring catwalk-inspired fashion to the high street, as quickly and as cheaply as possible. The pressures being placed on the factories to complete the orders on time will have come directly from the brands in question. However Rana Plaza is far from the only social disaster of recent times, from late 1990 to the present day there have been 28 reported incidents in garment factories with 22 of these having fatalities. During this time, almost 2000 factory workers have lost their lives due to various social compromises being placed on the manufacturing supply chain (Bhuiyan 2012).

The need for social responsibility in the supply chain has never been more prominent. Social responsibility can be defined as when all human interaction in the clothing supply chain work in good working conditions and are paid a fair living wage. This term, however, is often misunderstood and frequently interchanged with other terms such as ethical or sustainability. Whilst social responsibility and sustainability often come hand-in-hand, the definition between the two can be quite clear. Social responsibility and ethical refer to the human interaction within the garment supply chain, while sustainability is the long-term durability of the environment. A further issue when attempting to define this term is the lack of industry standard, leaving the meaning to be very subjective and interpreted very differently from company to company. As previously discussed, social aspects of the supply chain are not limited to just one stage of the process and can affect different people in many different ways. Social responsibility can refer to working hours, working conditions, health and safety of the working environment and worker's pay. It has been suggested that when discussing ethics the term is far too broad in its definition, too loose in its operations and too moral in any other stance (Devinney et al. 2010).

Despite the difficulties surrounding the terminology there are many examples of engagement from the perspective of the retailer. This again comes in many different forms from retailer to retailer, with many setting goals or targets to aim for in the near future. The use of more organic cotton, further engagement in Fairtrade and the use on non-toxic dyes are all generic examples of such engagement. Marks and Spencer can be provided as an industry example, where in their Plan A commitments they aim to work closely with their manufacturers, setting up educational training schemes for their workers to help them gain basic literacy skills and knowledge regarding health issues in order to better support their families. In addition to this, Marks and Spencer also works closely with manufacturing locations, contributing to town resources such as educational institutions and sanitary facilities. They have also worked on the development of several *green model factories,* the first of which was in Sri Lanka where huge quantities of trees were planted in the factory vicinity. This not only provides further jobs for local people, but also offsets the carbon emissions produced when manufacturing the fashion goods. Another example of retailer engagement would be where the retailer has engaged their customers in initiatives in the context of social responsibility. Recycling schemes based on the premise of consumers returning goods to store when they no longer want them has been launched by both H&M and Marks and

Spencer. Rewards in the form of money-off vouchers are helping to incentivise the initiatives. This consumer engagement will be discussed further later in the chapter.

This chapter will further explore the relationship between the consumer and the retailer in the context of the fashion industry. The exploration of this vital relationship will aid the reader in understanding the purchasing process and how the retailer has a great deal of power to influence the consumer in their choices. This is achieved through the utilisation of marketing methods and techniques and can often occur during what has been labelled in this chapter as the *window of opportunity* in retail. This is where the consumer has the intentions to buy a specific item but may be persuaded to change their mind, resulting in a different or increased purchasing behaviour. Recent changes in society can be held responsible for many changes in both consumer attitudes and the development of the value, lower end of the fashion market. The development of the fast fashion business model has increased pressure on the already long and complex supply chain, the consequences of which have been highlighted already. In addition to these pressures, several further challenges for the industry and in specific retailers will be discussed and the chapter concluding with how these challenges can be potentially addressed with recommendations for the future.

The chapter will begin by providing context and background surrounding social responsibility in fashion. This will guide the reader through the rationale for taking a more responsible approach to the production of fashion in the manufacturing supply chain. The purchasing process will be discussed in length in Sect. 2, highlighting the journey a consumer goes on in the lead up to purchasing a fashion product. This includes the internal decision making process undertaken alongside the influential factors which can adapt and change the final purchasing outcome. A case study conducted by the authors in 2013 has been included which explores the purchasing criteria of consumers looking in detail to the factors, which they consider to be most important when purchasing a product. The chapter continues in Sect. 3.1 and provides an insight into recent changes in the fashion market, discussing how economic and retail adaptations have a profound effect on the needs and wants of consumers in their fashion choices. The challenges facing social responsibility are outlined in Sect. 4.1, detailing the key factors currently preventing further change in the industry. The challenges are then addressed in Sect. 5.1, where potential solution and integral steps towards change are discussed. The chapter concludes with highlighting the innovative and powerful position of the retailer and how their role within this field can be the catalyst to moving the industry to a more socially responsible future.

2 The Purchasing Process

The purchasing process can be defined as a predetermined model which attempts to describe the actions of an individual on the lead up to and including the act of purchasing goods. A lot of models developed by authors are based heavily of the work of Ajzen (1985) in the development of the theory of planned behaviour. This

Fig. 1 Schiffman et al.
(2008) purchasing process

is where the act of behaviour is said to be a direct consequence of the theory of attitudes and where behavioural relationships are rationalised. Attitudes are developed within the context of subjective norms, perceived behavioural control, or intentions and behaviour in a casual fixed sequence. The initial development of Ajzen's model has led authors to adapt and modify this process within the context of their own work, resulting in the inclusion of other considerations such as implementation intentions, situational context and actual behaviour control.

The purchasing process as developed by Schiffman et al. (2008) can be described as a three-stage process, and occurs whenever a consumer interacts with a retailer to purchase goods. This process includes input, process and output (Fig. 1).

The first stage of this process is where consumers consider factors effecting the purchasing decision such as price, quality and item specification. The second stage is *process* which can be broken down into several sub-stages, the first of which is need recognition. This is where the consumer acknowledges that there is an internal need or want to purchase a specific item. Next, there is a competitor analysis, where the consumer carries out a search of the market in order to understand the scope of products that are available, which will fulfill the need previously recognised. Once the market has been analysed to a degree that the consumer is satisfied with (varying significantly from consumer to consumer), an informed purchasing decision will be made. The breadth of this competitor search will vary from person to person. Some will be satisfied with the first product found that matches the need criteria and is within budget and accessibility. Others will conduct a thorough search where a full range of possibilities will be explored, spanning an array of budgets and retailers. Some more savvy consumers will explore an even wider range of options, turning to online retailers such as Amazon and Ebay to try and find the best available product (often second-hand) for the best available price. The use of the Internet has increasingly made this competitor search easier and more accessible to the masses as consumers no longer have to physically visit retailers in stores to see the full product range available. This adaptation to retailer access has had a significant effect on the purchasing process in general as the behaviour of consumers is rapidly adapting to accommodate developments in technology such as advanced application of the internet. Once the purchasing decision has been reached, the final stage of this process concludes the product purchase. *Output* includes the consumer following on from their decision and both the act of physically purchasing the product decided on, and also the evaluation of that product once the consumer has received and utilised the product sufficiently.

The purchasing process has also been considered within a wider context, which has developed the purchasing process to move to that of a four-stage process. The first stage assesses the need for the product and could be compared to the input stage of the previous model discussed. Following this, the gathering of information must occur, which again can be compared to that of the first initial sub-category of

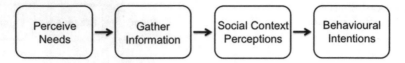

Fig. 2 Newholm and Shaw (2007) purchasing process

stage two of the process. The third stage referred to would be utilisation of perceptions of social context. This is where several authors believe the consideration for ethics and sustainability would come, an increasing consideration for the purchasing of many consumer goods. The final stage of the process which again is comparable to that of Schiffman et al. (2008) would be the act of developing behavioural intentions (Fig. 2).

This four-stage process as developed by Newholm and Shaw (2007) also believes that the large majority of decision making in purchasing occurs prior to the act of behavioural intentions, meaning that a consumer has made a decision before the physical action of purchasing.

2.1 Influencing Purchasing Behaviour

When considering the purchasing process as a three or four-stage model, there is an intervening period where retailers have the power to influence decisions made by consumers. This would be applicable in both an online and in-store situation where advertising, special offers and price promotions could influence the predetermined decisions made by consumers. This *window of opportunity* allows for the consumer's mind to be changed in preference for something cheaper or of a different specification. This point-of-sale marketing opportunity is usually utilised by retailers to upsell goods or to encourage consumers to over-consume and purchase additional goods, which were not included in their initial intentions. Retailers use marketing tools in order to do this, influencing the purchasing behaviour of their customers (Fig. 3).

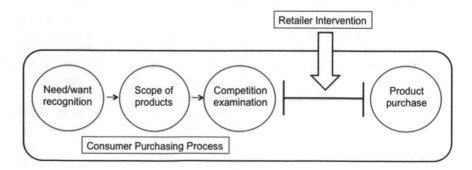

Fig. 3 Consumer purchasing process with retailer intervention (*Source* Authors)

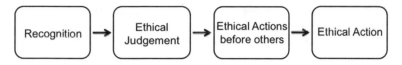

Fig. 4 Rest (1986) consumer purchasing process

This *window of opportunity* or retailer intervention has multiple applications on the part of the retailer, moving away from the upsell or over consumption element, to more of a social marketing approach. This is where retailers could use this window of power within the purchasing process to encourage consumers to purchase more responsibly, with the consideration of ethics and sustainability during the purchasing process. This period of time prior to the act of purchasing could be used to inform consumers further of ethical and sustainable issues pertinent to the field of consumer goods being purchased. This approach could not only help consumers to make more responsible choices, but to also become more informed and educated for future purchases made. When consumers apply such knowledge, it has been labelled as socially responsible purchasing as consumers are taking these factors into consideration prior to purchasing their choice of goods. This inclusion of ethical and sustainable considerations is said to be where the consumer morals and values regarding such issues will eventually have an effect on their final purchasing behaviour. The four stages are said to include recognition, application of ethical judgement, putting ethical actions before that of others and family and finally ethical action (Rest 1986). This model, (Fig. 4), implies that the utilisation of ethical actions occurs at a particular point of the purchasing process; however, it is also to be recognised that this can occur at many different stages. In particular at the end of use of the product, where the method of disposal may be considered taking into consideration recycling, up-cycling and biodegradable options.

When considering the purchasing process, there are a large number of possible factors that at many stages could influence the behaviour of consumers. These can be divided into two, by those that can be controlled by the consumer, called situational factors, and fixed factors which are out with the control of the consumer but still have the potential to influence behaviour. Situational factors encompass elements such as weather, personal finances and even mood of the consumer themselves. Fixed elements include factors such as price points, availability, store layout and design. All these elements have the potential to influence a consumer's purchasing behaviour during the purchasing process. Again many of these occur during the retailer's *window of opportunity*, especially the fixed influential factors that the consumer cannot control.

The influence of the price of a product is a heavily debated topic in literature when it comes to the discussion of purchasing behaviour. Ethics and sustainability acts as a catalyst to this discussion, as many consumers have to forsake responsible purchasing behaviour in favour of more affordable products. Cowe and Williams (2001) believe that price dominates the majority of decisions made by consumers and that there is regularly a trade off of ethics for improved price points. In addition to price it is believed that there are a number of other factors that consumers consider prior to buying an ethical brand, including; brand awareness, the product criteria and the convenience of purchasing an ethical product (Davies 2007). The prioritised list of considered elements on the part of a consumer has been named the purchasing hierarchy, with many authors again developing differing models of these influential factors. Price is thought to be the most important purchasing criteria, closely followed by value, quality and brand familiarity (Carrigan and Attalla 2001). Considering ethics and sustainability are not mentioned here, it can be assumed that consumers do not prioritise these factors.

Many attempts have been previously made to increase consumer awareness and knowledge of social responsibility through the application of initiatives and labelling campaigns (Bray et al. 2010). However, the development of too many labelling initiatives has consequently resulted in scepticism of the true meaning behind these labels on the part of the consumer. In addition to the meaning, the effectiveness has also been questioned, providing a certain amount of doubt on how useful, if at all, social labelling can be. This negativity is also thought to have spread towards the retailers and the brands behind the labels, with consumers holding them responsible for their initial scepticism. This common approach from consumers is thought to be having a negative affect on the ethical market, with Cowe and Williams (2001) believing that a generation of disinterested consumers could kill off the ethical movement.

The influence of the scepticism of ethical product labelling, however, does assume there to be two very distinct markets; that of socially responsible goods and one that is not. Again this assumes that these two markets are available to the consumer rather than just the one morally correct option (Niinimaki 2010). Whilst there remains to be two distinct markets, consumers will continue to have a choice to make during the purchasing process, whilst taking the more integrated approach as previously mentioned by Arnold (2009) would eliminate a choice situation. This scenario would see brand and retailers incorporate socially responsible values into their core business, resulting in a more ethical and sustainable offering to con-sumers. This approach, however, would very much rely on retailers and brands building trust with their consumers, where communication of their shared values would need to be clear and accessible to their customers. In an interview conducted with the head of responsible business from a leading UK high-street retailer, it was discussed that this was the ideal situation. Consumers could build relationships with retailers who share their moral values and trust them as a business to make the correct decisions for them. This was said to see the consumers leaving their worries at the store doorway, and consequently shopping without a choice or trade-off needed (James 2015).

A heavily debated issue within literature is the accessibility of socially responsible clothing to the mass market (Bray et al. 2010; Niinimaki 2010; Arnold 2009; Carrigan and Attalla 2001). This is thought to be one of the largest influences impeding the growth of the ethical and sustainable market. This lack of availability of such goods is further hindered with a lack of choice, meaning that not only are these goods in most cases unavailable, but when they are available the choice is very limited. This is particularly pertinent to the fashion market as consumer purchasing choices are predominantly based on aesthetics. Authors have gone as far as to believe that consumers have a limited interest in ethical and sustainable issues due to this lack of choice and availability (Niinimaki 2010). This again results in consumers having to make choices and often trade-offs between aesthetics and ethical or sustainable products, which mirrors the idea previously discussed where price can also result in an either-or situation (Cowe and Williams 2001). As a result debates surrounding this topic have moved on, suggesting that in order to be successful the fashion market needs to bring together trend-led collections and responsible values (Arnold 2009).

Socially responsible purchasing has often been related to compromise or trade-off situations, where consumers have to make a choice of one factor over another. The term *flexibility* has been used in the context of such purchasing behaviour, implying that socially responsible behaviour has to be balanced with everyday life practicalities. There is said to be a need for a balance between this practical approach and the application of ethical values (Szmigin et al. 2009). However, due to many conflicting influences as previously discussed, these balances often cannot be reached and the consumer develops a justification strategy. This is where consumers attempt to justify (with themselves and others) why their non-responsible behaviour is acceptable. This form of rationalisation compromises a consumer's own moral values yet permits the opposite behaviour despite them knowing otherwise. Consumers attach logic and meaning to their decision making at this point to allow them to conduct behaviour to compromise their knowledge and awareness of ethical and sustainable issues (Auger and Devinney 2007). This series of justifications has also been referred to as neutralisation, where consumers dilute their responsible behaviour through justification strategies and later deny all negative consequential impacts (Chatzidakis et al. 2007). This point is supported by Niinimaki (2010) who believes that consumers subconsciously make decisions based on their own individual needs. These decisions can affect and benefit those needs. This approach when taken by consumers can result in them feeling very unconnected with where their clothing comes from, having no awareness, knowledge or empathy of the manufacturing supply chain, let alone the knowledge of the ethical or sustainable issues that can occur.

A further issue which is having an impact on socially responsible behaviour is the feeling that a consumer's contribution is not enough to make a difference. This phenomenon entitled *perceived consumer effectiveness* (PCE) (Ellen 1994) refers to the level of affect that a consumer believes their contribution to be making. When a consumer believes their actions have little or no effect, this is termed low perceived effectiveness (LPCE). When a consumer feels this way it is believed to have a negative impact on their socially responsible intentions overall, which again when referring to Ajzen's Theory of Planned Behaviour, is said to be indicative of behaviour.

Culture is also thought to play a role within the consumer purchasing behaviour, especially within the context of ethics and sustainability (Belk et al. 2005).

Case Study—Consumer Purchasing Hierarchy

The authors conducted a study in October 2013 of 35 consumers who were asked to provide their purchasing hierarchy criteria. A controlled group of participants were targeted through an ethical fashion symposium organised by Fashioning an Ethical Industry organisation that aids students and tutors in fashion related courses. These participants were selected due to their existing interest in ethics in fashion, where their purchasing criteria could then be assessed. The participants were predominantly academics and students who had an interest in ethical issues surrounding the fashion industry. The majority of participants were enrolled on undergraduate courses from various universities and colleges in the UK. Consequently a large proportion of participants were aged between 15 and 24. However, there were a number of academics and tutors that participated who fell within the age bracket of 35–44 or 45+. This wide range of participants posed several interesting areas of inquiry, including if age and salary influenced the type of garments purchased and the retailer those items were bought from.

This study aimed to gain an insight into the most important considerations to consumers when purchasing fashion items. The study saw participants providing information of their top five considerations during their fashion purchasing process. They were provided with eight choices, being asked to rank their five top important factors in descending order from most important to least important. This approach along with the example choice answers were established from a previous preliminary piece of work conducted with a major high-street retailer. These included: handmade, organic, Fairtrade, price, aesthetics, locally sourced, material and washing instructions.

The results were discussed as a group but then recorded using a pre-prepared template, which included visual representation in sticker format (Fig. 5). This approach was taken to make the exercise less formal, prompting further discussion in the group.

In addition to ranking the most important purchasing criterion elements, participants were also asked to provide rationale for these choices. This allowed the author to begin to understand the reasoning for participant's decisions and how this ultimately influenced their purchasing behaviour.

The results from this study were analysed as a whole before this was then further broken down into each of the top five choices provided. For the overall summary of the study, the top three choices (in descending order) from the participants were aesthetics, materials and price (see Fig. 6). Locally sourced, Fairtrade and Organic all scored fairly low, indicating that these factors were not a priority to participants when purchasing fashion. Handmade was the lowest scoring factor, again indicating that this is of the lowest importance to participants, however this may be due to the participant being more familiar with mass produced garments.

When breaking this data down into more specific hierarchical choices made by the participants, there appeared to be patterns emerging in the data with a clear

Fig. 5 Recording template
(*Source* Authors)

divide indicating factors that were necessities and those which were just desirable. The top four choices which were described by participants as necessities were aesthetics, material, price and washing instructions in comparison to those much lower scoring elements which were described as desirable but not necessary; locally sourced, Fairtrade, Organic and handmade. The qualitative rational provided by the participants allowed this conclusion to be reached (refer to appendix for full data set).

Choice No. 1
The first hierarchical choice from the study was aesthetics, with 87 % of participants at this point indicating that this was the first thing they looked for when purchasing a garment. The remaining answers indicated that material was also a consideration at this point.

Choice No. 2
The second choice indicated that the material of a garment was the most popular answer at this stage with 53 % of participants indicating that this was the second

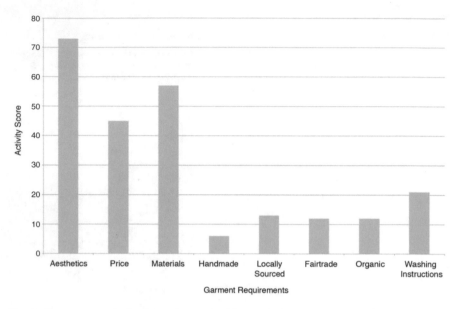

Fig. 6 Consumer purchasing hierarchy results (*Source* Authors)

factor they considered during their purchasing process. The rationale for this answer predominantly focused on material being an indication of quality and that participants were looking for a reflection of quality in their clothing purchasing decisions. Other answers at this stage were price with 33 % of the participants indicating this.

Choice No. 3
The third choice made by participants was price with 46 % of participants indicating this. However during this hierarchical choice, material was also mentioned, with 40 % of the votes. This small differential percentage indicates that these two elements are both essential choices during their initial decision making process.

Choice No. 4
The fourth choice in the study begins to really show where the divide between necessity and desirable begins. The majority of participants at this stage had indicated that price, aesthetics and material were in some order within their top three choices, with this now being the stage where people may start moving towards considering non-essential criterion within their purchasing process. This choice saw the widest spread of votes, however there was one answer that dominated the category, which was washing instructions. During the participant rationale this appeared as again an important factor to consumers during the purchasing process, with many indicating the importance of looking after the clothing they buy.

Choice No. 5
The final choice within the study, choice five saw a plethora of desirable choices being made, again highlighting this clear divide of necessary and desirable factors.

Fairtrade received the most attention at this stage with 27 % of participants indicating this. However there was again a wide scope of answers, with the focus being very much on hand made, locally sourced, Fairtrade and Organic.

This study showed the purchasing criteria of the participants, indicating the five most important criteria to them when making fashion choices. The results indicate this clear divide between necessary and desirable which shows that ethical and sustainable issues are rarely a consideration until after the essential criteria have been met. This reflects Maslow's Hierarchy of Needs, where consumers must have reached the previous stages of the pyramid prior to self-actualisation (at the top of the pyramid) where at that point social responsibility may be included in their purchasing hierarchy. The rationale behind participant choices were interesting and aided in conclusions being reached regarding their purchasing hierarchy.

2.2 The Intention-Behaviour Gap

In recent years, research in the area of socially responsible attitudes and behaviours has lead to the identification of a distinct disparity in consumer ethical intentions translating into actual behaviour. This has been labelled the intention-behaviour gap and has been the focus of many studies in the past 10 years (Ozcaglar-Toulouse et al. 2006; Bray et al. 2010; Cowe and Williams 2001; Worcester and Dawkins 2005; Belk et al. 2005; Auger and Devinney 2007; Carrington et al. 2010). This gap has also been 30:3 syndrome (Cowe and Williams 2001), which refers to the numeric figures that initially lead to the identification of the intention-behaviour gap. The statistics supporting this phenomenon indicate that 30 % of consumers have the initial intention to purchase responsibly. However when this is to translate into behavior, only 3 % of the original group of consumers actually purchase responsibly. This indicates that intentions cannot be relied upon when it comes to socially responsible purchasing behaviour, going against theory developed by Ajzen (1985), which states that intentions translate into behaviour (Fig. 7).

This disparity from intention to behaviour has been rationalised by many authors who believe that societal issues such as social desirability may be accountable (Worcester and Dawkins 2005). This is said to be one of the reasons why consumer intentions far outreach that of their consequential behaviour and occurs during the execution of the research methods utilised to collect the statistical consumer data. It is thought that consumers are offering answers more socially acceptable during these data collection exercises, which has been found to be a reoccurring issue within socially responsible research conducted with consumers. Researchers in the field are working to overcome such issues that in time should result in more accurate data for the relationship between ethical intentions and behaviour (Auger and Devinney 2007; Dickson 2013). Further statistical research conducted in the field does confirm there to be an intention-behaviour gap (Worcester and Dawkins 2005), however when considering the purchasing process the *window of*

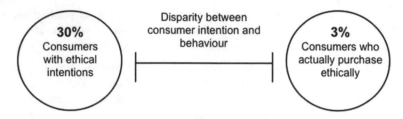

Fig. 7 The intention-behaviour gap (*Source* Authors)

opportunity, as previously discussed, must also be considered. This is where consumers may maintain their socially responsible purchasing intentions up until the point where they physically go shopping, where other factors such as price, aesthetics and availability may cause a choice or trade-off to be made.

Researchers have continued their work in this field to investigate why the intention-behaviour gap occurs with some suggestions including that social responsibility is not considered by consumers during their purchasing decision criteria (Carrigan and Attalla 2001) and others including the consumer not being knowledgeable of such issues in order to make an informed decision. It has also been suggested that the lack of obligation to engage with ethical and environmental issues is also accountable for the gap between intentions and behaviours (Ozcaglar-Toulouse et al. 2006). Other authors, however, blame issues previously discussed such as the lack of goods provision and that available product is not trend orientated or desirable.

3   Changes in Purchasing Behaviour

3.1   Market Development

When the recession hit the UK in 2008, significant changes in not only the way people purchase fashion but also the provision in the fashion sector were seen. Consumers were used to having the option to purchase fashion, however, with financial changes in society came with it cuts in people's disposable income. Fashion for many people is a hobby or a pastime, where recreational spare time will be spent perusing stores on a Saturday afternoon with friends. For a lot of people it is more about the physical act of shopping rather than the outcome of their shopping trip. This downtime is utilised by consumers to relax, see friends and enjoy themselves, and if they purchase fashion goods this is seen as a bonus. Retailers are also responding to this, applying more money, time and effort into the experience of shopping for their customers. The fashion brand Hollister is a great example of this. This retailer heavily controls the shopping experience that their customers have, influencing key experiential factors such as light, sound and even smell. The lighting is kept minimal and could even be described as dark, with clothing often being highlighted by spot lamps. The music reflects heavily of the brand and is chosen very carefully. The smell experienced during shopping is also very unique; the same scent is pumped out in all stores to begin to build brand recognition with their customers. The consistency through their stores is also carefully monitored, with the experience in each store being identically replicated. This is extended as far as similar looking staff, a carefully created brand that heavily controls the retail experience of consumers. This store has a cult following of consumers who want to buy into the brand and often visit knowing full well they will not purchase but simply window shop and absorb the carefully crafted brand marketing.

As a result of the recession, consumers felt they could no longer shop as they used to prior to the economic downturn, which saw a dramatic fall in sales and consequently profits for the vast majority of the fashion high street. This resulted in an adapted model emerging that allowed for consumers to still be able to shop in the way they had become accustomed but without the same price tags. This resulted in the emergence of the value market sector, which as a direct response to the change in consumer purchasing habits, grew exponentially in a short space of time. At the time, sales for the value end of the market saw a growth of 6 % taking its value up to £8.1 billion with research showing that 36 % of consumers then favouring the value end of the market (Mintel 2009). This growth also saw consumer preference for quantity over quality, with not only purchasing habits moving towards the value market sector but also the quantity of value goods being purchased increasing, again as a reflection of price. This saw the high-end, luxury end of the fashion market suffer with purchasing behaviour indicating a preference to fashion retailers who could deliver regular up to date trends, whilst still remaining competitive on price.

The high street in general began to change the type of goods being offered to consumers with the emergence of many more value stores and pound stores becoming a new phenomenon. Consumer preference to these types of stores continued to grow with many people favouring pound stores in preference to special offers offered by the more traditional stores (Portas 2011). Many consumers are now favouring the value end of the market (Ritch and Schroder 2009) with a large increase in the provision of pound stores and value retailers. It is this bargain-hunting hunger that allowed consumers to feel they were still purchasing goods (often larger quantities of cheaper goods) despite the effect the recession was having on their financial situation (James 2015). The increase of cheaper priced goods, however, has had a very negative impact when it comes to the value a person places on a purchased item. This has been heavily evidenced with fashion, with consumers buying fashion items sometimes to only be worn once, seeing them engaging in a throwaway society. Another result of the popularity of the value end of the market is more traditional stores began to offer a wide range of special offers and value buys. However, an increase in these offers such as BOGOF (buy-one-get-one-free) are said to have encouraged inflation in original prices and encourage over consumption purchasing activities (Arnold 2009). The recession also saw an increase in the use of everyday credit, which again facilitated consumers to continue their purchasing habits, utilising credit and store cards to help facilitate this. This use of credit has also had a negative impact on the way people value products they buy with this being due to no longer having to patiently save money in order to buy their desired goods. However it is stores which offer brands at a reduced price, such as TkMaxx, which are said to be offering consumers the most value, in terms of quality, for their money. This approach to retail is possible due to the retailer purchasing unsold stock from other brands very cheaply that allows them to offer an appealing price point to consumers.

Since the recession, studies have shown that there has been a direct influence on the purchasing of ethical and sustainable products (Worcester and Dawkins 2005), with consumers replacing ethical consideration with those closer to home. Arnold (2009) believes that consumers start with considerations in their immediate family circle, progressing to that of friends and others close to them. Again this directly reflects Maslow's theory, where consumers start basic in their behaviours, progressing gradually and eventually considering additional factors such as ethics and sustainability. It has been shown that a small sector of better educated individuals are moving towards the top of Maslow's pyramid and incorporating responsible criteria into their decision making process (Soloman and Rabolt 2004). This sees consumers wanting something additional to consider, something outside of themselves and their immediate circle, something precious and almost spiritual (James 2015). This sector of people begin to start trying to be the best version of themselves they can be and some achieve this through the consideration of ethical and sustainable issues within their purchasing behaviour.

In addition to the growth of the value sector, there have been many more changes identified on the generic shopping high street. For example the value sector has extended beyond the high street to supermarkets now offering a wide range of womens, mens and childrens wear. The three leading supermarkets (Tesco, Sainsbury's and Asda) offer clothing to their consumers out of convenience, appealing to the needs and wants of their consumer profile. With every £1 spent on shopping, 50 pence of that is being spent of food and groceries in supermarkets (Portas 2011). Convenience is a real driving force for supermarkets, as they begin to continuously extend their services and concessional stores to accommodate the growing busy lifestyle of their customers. These services include hairdressers, cobblers, bureau de change, banks, cafes and sunbeds. Supermarkets however are not the only stores, which have adapted and emerged to facilitate shopping convenience. The market has also seen a large growth in the development of out-of-town retail parks, which again facilitate an element of convenience for consumers. The parks offer a wide range of shop types, from furniture to fashion and food to beauty. In addition to stores other services are also available such as cinemas and restaurants which again goes towards building a shopping experience based on convenience for their customers. Out-of-town retail parks also offer free parking and due to geographical location often facilitate consumers avoiding city centre traffic and congestion charges. This approach has been named need-based retailing and only highlights further that in-town shopping high streets are not keeping up with this development (Portas 2011).

The number of high-street stores has fallen dramatically, with a decline of almost 15,000 stores in the period 2000–2009. With an additional 10,000 losses predicted for the coming years, approximately one in six shops stands empty (Genecon LLP and Partners 2011). Retailers have recognised these changes and downturn of the traditional shopping high street with some brands such as Topshop, now deciding to expand overseas markets in preference to the UK (Portas 2011).

The change in consumer needs has also been reflected in the seasonal fashion cycle, seeing this tradition cycle increase in speed and ultimately speed and quantity of the goods being on offer to consumers. Moving away from the traditional two-season approach, the fashion market is accustomed to a constant drip-feed effect of 14 rolling collections in any one year. This constant delivery approach will ensure consumers stay up to trend on their fashion purchases, however, has hugely negative impacts in the increased level of consumption for example. This vast increase in the volume of goods delivered to store and available to consumers each year is not exclusive to the lower end of the fashion market. There has been an increase of pressure on the high-end, luxury sector of the market, with designers intensifying their offering through the addition of pre-collections, again doubling their traditional offering. This move has incorporated the development of transitional seasons and has seen designers such as Stella McCartney incorporate Pure Summer and Pure Winter to her collection each year. This adapted approach, however, has been defended with it being described as one collection, with only the delivery changing, adopting this constant drip-feed of trend-led fashion available to the market (WGSN 2010).

The series of changes identified and discussed indicates not only a change in consumer purchasing behaviour, but as a direct consequence, a change in the offering provided by brands and retailers. This constant need for new and better things has been acknowledged and reflected in a constant drip-feed effect of new designs being delivered to store. With the recession as a catalyst for this change in behaviour, the high street needs to keep up with the volume and dexterity of the changes, adapting with their consumers in order to remain a key feature of the fashion market.

3.1.1 Fast Fashion

These changes in purchasing behaviour and the increased demand for constant new designs and product offerings have facilitated the emergence of the fast fashion business model. Fast fashion can be defined as bringing catwalk-inspired fashion to the masses, as quickly and as cheaply as possible. Fletcher (2008) described this as resulting in a change in fabric and manufacturing quality of clothing, with social or environmental compromise in the supply chain. A fast fashion garment is made to the standard of quality to be worn only 10 times before being disposed of, which again is indicative of the speed of consumption as a reflection of the speed of new designs being delivered to store.

The garment manufacturing supply chain is often a long and complex process, with more often than not several different countries in the world being involved from fibre to garment. Despite the length and complexity of the supply chain, the speed as a result of the fast fashion business model has significantly picked up pace, with the speed from start to finish dramatically increasing. This increased speed of delivery, however, has not been acknowledged in the supply chain, with no innovative adaptations being applied to facilitate this change. In order to meet the demanding lead times of the fast fashion supply chain, compromise of this kind is often required (Fletcher 2008). This can often result in social disasters such as the Rana Plaza disaster previously discussed in the chapter. However, social compromise comes in many forms and does not have to result in such a disaster in order to constitute social pressures. Long working hours, unpaid overtime, non freedom of association, forced labour, poor working conditions and not being paid a living wage are but a few examples which are regularly witnessed when social compromise occurs in the manufacturing supply chain. Stuart Rose, former Chief-Executive of Marks and Spencer highlights the importance of retail price in the context of the supply chain; 'how can you sell a t-shirt for £2, and pay the rents and pay the rates and pay the buyer and pay the poor girl of boy who is making a living wage, you cant'. This pertinent point summarises the crux of social compromise that can occur during the manufacturing of fashion. Heightened media coverage has lead to the development of the term sweatshop, which is now widely used to describe the unsatisfactory working conditions endured by over 100 million garment workers worldwide (Lee 2007). This unsustainable business model has been summarised by Hawkins (2006) who reflects on the current consumption level in the context of the environment stating that if the consumption levels are to continue in the developed world, the output (natural resources) of three planet earths are needed.

In addition to social issues, environmental impacts are also rife in the fashion business model. The most prominent example of this as a direct consequence of the fashion industry would be the disappearance of the Aral Sea. Geographically located in Uzbekistan, a region which is rich in cotton growing, utilised the Aral Sea for many years in order to grow this water intensive crop. However, an increase in farming increased the need for water and has lead to the disappearance of the water source all together, leaving dry, desert land where the sea once was. Between 1960 and 1989, its area decreased by 40 % demonstrating the rapid decline of one of the worlds four large lakes (Giesen 2008). Other sustainable issues occurring in the fashion supply chain are carbon emissions, overuse of natural resources (water and oil.), water contamination and also post-consumer issues such as non-biodegradable fibres being sent to landfill.

The current high-street leaders of fast fashion are Primark who have in the past been related to the production of garments in the Rana Plaza factory prior to its collapse (Nelson and Bergman 2013). Whilst this disaster, happening in 2013, was expected to have negative impacts on their sales and profits due to the media attention received by Rana Plaza, reported profits in November 2013 were up by 44 % on the previous year. This saw profits sore to £514 billion for that year alone

(Hawkes 2013). Whilst this could have been a consequence of increased press coverage, albeit negative, the fast fashion model adapts itself to the retailer collecting a higher net margin of the overall retail sale value. This is heavily due to fewer pieces being delivered to market at a higher speed, promoting increased consumption (Tokatli 2007). Ritch and Schroeder (2009) compare this fast fashion model of consumption to McDonalds; cheap, fast, mass produced, hassle free and reliant on social and or environmental compromise.

3.1.2 Purchasing Motivations

As previously discussed, the purchasing process begins with the need or want of recognition of a new product or service. Whether this is a want or a need, something prior to this decision motivates consumers to acknowledge they need something new. This motivation remains with the consumer throughout the purchasing process helping them to make decisions at every stage. As mentioned previously, Maslow who was an economist formalised consumer motivations in his Hierarchy of Needs theory. Pyramid shaped, the premise that consumers move from the bottom to the top, only progressing to the next stage when the previous has been achieved. The five levels of the pyramid include; *physiological*; referring to the most basic level of requirements including food, water, and sleep, *safety*; taking the requirements to the next level including the security of family, finances and employment, *love/belonging*; expressing the need for others including friendship and intimacy, *esteem*; which sees the need for interaction and respect from others and *self-actualisation*; including morality, creativity and problem solving (Fig. 8).

This basic theory can also be applied to the requirements of clothing for consumers, taking the principals originally developed by Maslow and recontextualising these for the fashion market. In ascending order:

- *Physiological;* the basic need for clothing in its simplest form, the need for clothing to cover the human body for modesty.
- *Safety;* the functional need from clothing including security and protection from the elements—warmth, cool, etc.
- *Love/belonging;* the purchasing of a certain type of clothing in order to fit within the desired consumer tribe
- *Esteem;* the acknowledged respect from peers in reference to appearance, mutual forms of respect amongst peer groups
- *Self-Actualisation;* the need to progress to the next level and problem solve within fashion, to give something back to society.

It is only when reaching the peak of the pyramid that ethics and sustainability are considered, meaning that prior to this stage, this is not a consideration in consumer requirements from fashion. This pyramid type approach can also be adapted to the purchasing process, where consumers move through the stages in the steps leading to the act of purchasing a product.

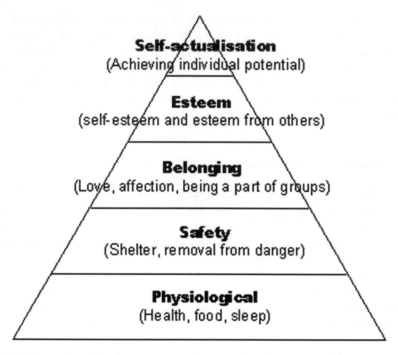

Fig. 8 Maslow's hierarchy of needs

As previously discussed, the purchasing process is a basic three-step process: input, process and output. However, when considering existing literature and the development of the process in the context of changing consumer behaviour, this can be progressed to the following five stages leading to purchasing behaviour. The first stage of the process (in ascending order) would be need *recognition*; identifying the need or want for a product of service, *information search*; the analysis of the product market, identifying products which will satisfy the need or want, *competition evaluation*; the scope of the competition to find the best product available at the best price, *ethical and moral values*; the opportunity for consumers nearing the peak of the pyramid to consider social responsibility in their purchasing decisions made, *situational attributes*; taking into consideration the precise situation in the context of marketing strategies (fixed influential factors) and situational elements such as weather and mood.

This comparison between the requirements of fashion and the process leading up to the point of socially responsible fashion purchasing has been demonstrated in the ethical purchasing hour glass, as developed by the author in 2012 (Fig. 9). This demonstrates both the purchasing behaviour hierarchy of consumers and also the product to purchase selection process, both of which, once all stages have been achieved, can lead to socially responsible fashion purchasing.

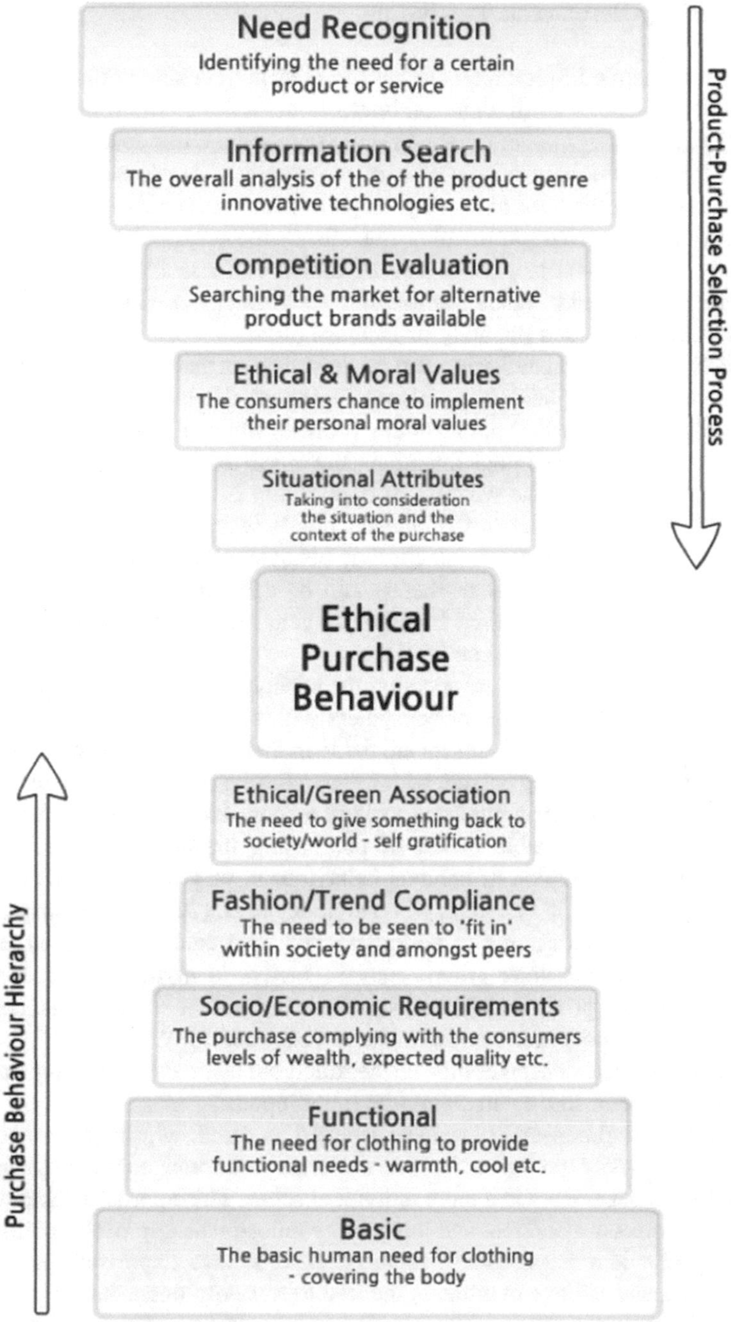

Need Recognition
Identifying the need for a certain
product or service

Information Search
The overall analysis of the of the product genre
innovative technologies etc.

Competition Evaluation
Searching the market for alternative
product brands available

Ethical & Moral Values
The consumers chance to implement
their personal moral values

Situational Attributes
Taking into consideration
the situation and the
context of the purchase

Product-Purchase Selection Process

**Ethical
Purchase
Behaviour**

Ethical/Green Association
The need to give something back to
society/world - self gratification

Fashion/Trend Compliance
The need to be seen to 'fit in'
within society and amongst peers

Socio/Economic Requirements
The purchase complying with the consumers
levels of wealth, expected quality etc.

Functional
The need for clothing to provide
functional needs - warmth, cool etc.

Basic
The basic human need for clothing
- covering the body

Purchase Behaviour Hierarchy

Fig. 9 The ethical purchasing hourglass (*Source* Authors)

3.1.3   Socially Responsible Purchasing

Socially responsible fashion refers to the social or human side of the supply chain, and can be defined as when all social interactions in the clothing supply chain are paid a fair living wage under good working conditions. Often referred to as ethical fashion, this terminology brings with it complexities due to a lack of industry standard, meaning the term to be very subjective and interpretive. This irregularity and often confusion over terminology results in the term being interpreted very differently from person to person. The execution of these ill-defined morals also results in retailers and brands interpreting such values in different ways, often utilised to facilitate good publicity or increase profit.

Socially responsible purchasing can be described as the consideration of ethical or social values and morals when undertaking the purchasing process. These values can be considered at various points of this process, whether it is prior to the product search and applied after initial intentions but prior to the purchasing behaviour. Again referring back to the window of opportunity in the purchasing process, as previously discussed earlier in the chapter, this is where retailer and brand could ultimately influence consumers purchasing behaviour through the application of social marketing tools. Social marketing can be defined as utilising the common tools of marketing, being utilised to influence behaviour for the good of society or the environment (Kotler and Lee 2005).

Consumers who engage in socially responsible fashion purchasing can be described as ethical consumers or consumers with a conscious. It has been widely recognised, however, that consumers are fickle in their purchasing behaviour, which may change each time they engage with the purchasing process. Consequently, this means that a consumer may purchase socially responsible clothing one day and not the next. This unpredictable nature of purchasing means that a predication of behaviour or the discussion of habitual behaviour is very difficult. This has again been widely discussed in literature, with Devinney et al. (2010) expressing the need for the term ethical consumer to be approached with caution, going as far as to believe that these consumer criteria may in fact be a myth. This is due to an idealistic consumer profile being created, where the perfect ethical consumer is put forward as a role model for consumers to be compared to. When doing so, certain elements of this ideal consumer may be identified but again due to the unpredictable nature of behaviour, this again cannot be relied upon.

The term ethical consumer has been labelled as dated, with more recent developments in the field preferring to use the term *consumer social responsibility*, which refers to consumers in a more individual sense. This approach acknowledges that all consumers are different and that a more tailor-made approach to influencing consumer behaviour is necessary. Industry experts have expressed the need for individualism and believe that this is the key to not only understanding consumer purchasing behaviour but to also then know how to influence it (Barrie 2009). Despite this individual approach appearing to be the most logical due to the unpredictable nature of purchasing behaviour, the use of consumer typologies or tribes are widely acknowledged by academia and industry alike (Clouder and

Harrison 2005; Cowe and Williams 2001; Szmigin et al. 2009; Carrigan and Attalla 2001; Morgan and Birtwhistle 2009; Mintel 2007). However when analysing these consumer typologies further, the relationship between socially responsible fashion purchasing and demographics is inconsistent and not indicative of any future behaviour (Devinney et al. 2010). Just as with the purchasing process, consumer typologies are also reflected differently according to different research, brands or retailers. For example, Clouder and Harrison (2005) believe there to be three segments of consumers within the context of ethics and sustainability; distancing, integrated and rationalising.

Another example of this methodology is British high-street clothing brand who believe that there are five different consumer categories within a social responsibility arena. During an interview with the head of sustainable business, it was detailed that there are three initial categories of consumers; 70 % of which could be described as *average,* 20 % who are said to have *no care of consideration* for ethical and sustainable issues and the remaining 10 % could be described as *green consumers.* Through further conversation, there were two further categories discussed which were; *don't believe they can help* and *willing but don't know how.* The interviewee went on to discuss that it would be small steps to progress, moving towards the small percentage of *green consumers* that would aid the industry in moving to a more sustainable future (James 2015).

This approach considers the broad spectrum of consumers and their attitudes towards the engagement with social responsibility. Several sections as described by this brand can be related back to previous points made earlier in the chapter. For example, *don't believe they can help* relates back to the theory of perceived consumer effectiveness developed by Ellen (1994) where consumers consider the level to which their contribution to socially responsible purchasing contributes to the wider cause. *Low perceived effectiveness* describes when consumers feel that their contribution will help very little, the very notion that is acknowledged by the British, high-street clothing retailer previously discussed. Another example of this would be *willing but don't know how* which has a direct relationship with the impact consumer knowledge and awareness has on their socially responsible purchasing behaviour. Non-ethical behaviour has on many occasions been related to a consumer lack of knowledge of ethics and sustainability in the context of fashion.

The idea of an informed decision making process is again a heavily debated topic, again relating to the relationship between knowledge and behaviour. An informed consumer could be described as someone who possess' an adequate amount of knowledge in order to make an informed decision regarding their purchasing behaviour. However, again the subjective nature of social responsibility means that levels of knowledge will vary considerably and again this will have an impact on decisions made. The definition alone of social responsibility is interpreted by people very differently, let alone the way they relate this topic to their fashion purchasing decisions made. It is also to be acknowledged that due to the complexities with consumer knowledge and awareness levels that again the ideal informed consumer may be unattainable. This does not, however, discredit the importance of purchasing fashion consumers having a good level of knowledge of ethical and sustainable issues pertinent to the fashion industry.

Knowledge and awareness levels of such issues do, however, rely on consumers implementing this during their purchasing behaviour which is often seen not to be the case. This again questions the relationship between a consumer's knowledge of social responsibility and their actual behaviour. Relating back to the intention-behaviour gap, not only external factors can intervene in the initial intentions translating into behaviour. This gap could also be contributed to by consumers themselves and be affected by internal decisions they make when purchasing fashion. For example, they may be fully aware of social and environmental issues relating to the clothing supply chain, however, they may choose not to use this knowledge in their purchasing decisions. This refers to what have been labelled *justification strategies*, where consumers consciously rationalise their unethical behaviour for one reason or another. For example 52 % of consumers in the UK claim to be ethically aware, but admit to not purchasing as a reflection of this knowledge (Worcester and Dawkins 2005). This statistic alone confirms that it could be the consumers themselves causing the intention-behaviour gap to exist. However, to consider this from a different perspective, there could also be consumers that do not intend initially to purchase responsibly but, however, end up doing so. This could be due to a variety of subjective reasons just as with the generic consumer purchasing hierarchy; however, it could also be due to the implementation of social marketing strategies on the part of the retailers. This again could see an influence of purchasing behaviour during the *window of opportunity* as seen in the purchasing process. As discussed earlier, whilst this is currently being utilised to encourage further sales and higher profits, retailers also have the opportunity to influence purchasing behaviour for the better. Through the utilisation of social marketing strategies, retailers could influence a consumer to follow up their non-socially responsible intentions with socially responsible behaviour. This could create a further intention-behaviour gap, however this time it would be a development of non-responsible intentions to responsible behaviour.

4 The Challenges Facing Social Responsibility

4.1 Key Issues Preventing Change

Throughout this chapter, there have been many issues raised as having a potential negative impact on the development of social responsibility in fashion. Marketing and situational attributes have been discussed in the context of influencing purchasing behaviour, however a period of retailer intervention as identified during this process is where behaviour can potentially be changed for the better. Referring back to Fig. 3, which demonstrates the opportunity during the purchasing process where the retailer can influence a consumer's behaviour was discussed in terms of the purchasing choice moving from that of ethical intentions to non-ethical behaviour. This also relates back to the intention-behaviour gap which states that 30 % of consumers intend to purchase with the consideration of social responsibility, however when translating this initial intention into behaviour, only 3 % of consumers follow up their intentions (Bray et al. 2010). However this theory can also work in reverse, where the consumer may be influenced by the retailer during the window of opportunity in the purchasing process to consider social responsibility where they have not previously. This window of opportunity could be utilised by retailers to not only drive up sales and consequently profit through clever marketing strategies, but to also inform and educate the consumer in socially responsible issues affecting the products they are about to purchase. There are multiple benefits identified for taking this approach, which include;

- The increase of knowledge and awareness on the part of the consumer
- Improved levels of socially responsible purchasing
- Improved brand trust as the consumers are seeing the retailer to be responsible and active in the execution of their corporate social responsibility values
- The consumer will be further informed in order to make an informed decision regarding their purchasing choices
- The promotion of further repeat behaviour

Whilst approaches such as this could help improve the development of social responsibility in fashion, there remain many further issues, which are currently preventing change.

Consumer knowledge and awareness of ethical and sustainable issues remains one of the biggest issues currently affecting the sector, with many mass market consumers still very unaware of the negative implications of fashion production. Without this initial knowledge or awareness of these issues, consumers cannot choose if to engage or to implement this knowledge during their decision making process. If consumers do possess this knowledge they can then use it to inform their purchasing behaviour making adapted decisions in light of being informed. However the issue of implementation is also a significant one, with some consumers already possessing such knowledge but choosing not to use it to inform their purchasing process. This is where many trade-offs are made for either price or aesthetics, with the consumer implementing justification strategies to rationalise their non-responsible behaviour.

Despite there being many positive drivers to encourage positive behaviour, there remains very little incentives for consumers to engage in such behaviour. This again can be identified as an issue inhibiting further socially responsible fashion purchasing. The rationale behind a consumer engaging with ethics and sustainability in their fashion choices remains an issue with philanthropy being said to be the key driver to encourage this. Philanthropy can be defined as the love of humanity, in the sense of caring, developing and enhancing what it means to be human. This again relates back Maslow's Hierarchy of Needs, where people progress to the peak of the pyramid and seek to give something back to society in an almost spiritual way.

However this need alone cannot be the only incentive for a consumer to engage in social responsibility, and therefore further initiatives need to be implemented. An example of a retailer who currently recognises this would be Marks and Spencer who utilise 'thank you' campaigns to reward customers for their engagement. This usually involves an email campaign, social media coverage and promotional information on their Plan A website. The content for this campaign usually includes bold statistics that indicates to the customer the impact their contribution has helped towards. This again is related to the theory of perceived effectiveness (Ellen 1994) previously discussed where consumers think that their contribution is insignificant in the wider picture. In addition to thanking the customers that have engaged in the specific scheme, the campaign also sets out to highlight the impact of their contribution to others, encouraging not only repeat behaviour next time, but also new responsible behaviour in other customers.

In addition to the thank you campaigns, Marks and Spencer also utilise other initiatives to further engage their consumers in sharing their ethical and sustainable goals. The Swhopping scheme first trialed in-store in 2012 is where customers are encouraged to return their unwanted clothes back to store in order to be resold or recycled by Oxfam. This collaboration with a charity is a value set out by the retailer to help raise money and facilitate international demand and need for certain types of clothing. To date, 7.8 million garments have been *Swhopped*, worth an estimated £5.5 million, which has been donated to Oxfam (Marks and Spencer 2015). Encouragement for engagement in this initiative is the exchange of these clothes for a voucher to be spent in-store. This is naturally a favourable incentive for customers, however, in terms of social responsibility it could be criticised that this approach promotes further consumption of clothing.

Another challenge facing the sector would be the offering of additional ranges of more responsible clothing alongside main collections. This approach taken by many retailers does highlight the ethical or sustainable credentials to the customer, which could be seen as a positive. However when positioned alongside the core product offered by the retailer, it does reinforce that this is the exception and not the normal values utilised to produce all garments. An example of this would be a range of basics, which utilises Fairtrade cotton, being positioned next to those, which use non-Fairtrade cotton. This presents the consumer with a choice to make and encourages trade-offs and justification strategies to be implemented. When presented with a choice, the customer will turn to important factors such as cost, aesthetics and quality in preference to the ethical and sustainable values of the product. An example retailer who utilises this approach would be H&M who offer a small collection entitled *Eco-conscious*, which uses a percentage of recycled polyester in the production of materials for the collection. Another negative aspect of this approach is that it also promotes the segregation of social responsibility and encourages a separate market. The existence of two markets will not only continue to offer consumers the element of choice but also discourage retailers from embedding social responsibility into their core business values. As a result this approach also discourages the industry from moving towards a more socially responsible future, where the negative implications of fashion production can be developed.

During many studies with consumers regarding their engagement with social responsibility in their fashion purchasing behaviour, the implications of accessibility is highlighted, with many consumers not knowing where to buy products with such values. This is again a significant issue in the implementation of social responsibility with consumers feeling that they do not have the choice to buy these types of products even if their intentions are to do so. The high street facilitates the mass market access to fashion products and is also the lowest provider of ethical or sustainable products. Although these should not be offered as alternatives, the consumer remains unaware of what options are available to them if they wish to purchase responsibly. Retailers should be taking an integrated business approach to social and environmental values, which should again be communicated effectively to consumers who can then make their choice of retailer in preference to the compliant or non-compliant product.

A further issue frequently raised by consumers regarding the provision of ethical and sustainable goods is poor aesthetics. Ethics and sustainability has a historical association with being unfashionable and not on-trend, which again has implications when it comes to consumer's decision making process. The long-associated stigmas of ethical and sustainable fashion remain an issue with the quality, comfort and fit of such products also being questioned. Ninnimaki (2010) reiterates that ethical clothing is often not trend focused enough which could be putting people off engaging with such issues and ultimately affecting their purchasing behaviour. This again relates back to consumers providing justification strategies to rationalise their potential non-responsible purchasing decisions, with reasons such as not being on-trend being provided. In reality whilst ethical and sustainable products may not be widely available on the UK high street, there are plenty of interesting and on-trend boutiques and smaller brands engaging in social responsibility. This variation in business model, however, may have price implications, which could again put the average consumer off. A lack of desirable aesthetics could also cause consumers to again make trade-off decisions when purchasing fashion. This relates back to an earlier discussion regarding compromises having to be made in order for socially responsible purchasing to take place.

To summarise, there are several key elements as discussed that are currently preventing the fashion industry moving to a more socially responsible future, these are as follows:

- A distinct lack of consumer knowledge and awareness of ethical and sustainable issues in the context of the manufacture of fashion products
- The lack of implementation of existing knowledge of ethical and sustainable issues when engaging in the fashion purchasing process
- A lack of incentive or rationale for consumers to engage in socially responsible fashion purchasing behaviour

- Mixed messages being communicated by designers and retailers when offering separate/alternative collections that are more responsible than main line collections
- A lack of mass market accessibility to socially responsible fashion products
- A lack of availability of socially responsible fashion products that remain on-trend and fashionable

In addition to challenges facing the industry in the face of change, there have been several research issues found to be weakening, and in some cases invalidating data supporting several key arguments. For example, when assessing the market over-inflated research results indicating that more consumers are concerned with social responsibility than actually are. Weak research methods have been left accountable for this over-inflation of results. This methodological weakness has also been held accountable for creating inflated intentions in consumers, resulting in data that is inaccurate and not reflective of real consumer opinion. This relates back to the intention-behaviour gap where an over-inflated consumer intention could also be responsible for the disparity in behaviour translation.

4.1.1 Why Is Change Needed?

As discussed at the beginning of this chapter, there have been many recent events that have witnessed social and environmental compromise in the production of fashion products. The varying significance of these can be noted, however, the continuation of such consequential actions cannot carry on. The need for change within the industry has never been more relevant, with both consumer and retailer awareness on the rise, the time for the industry to start making positive changes is imminent.

The development of the fast fashion business model has played a key role in the negative consequences of the social and environmental compromise in the fashion supply chain. Intense pressure has been placed on the garment supply chain to deliver huge quantities of garments at a very quick speed. It is the negative consequences of this required speed that makes the fast fashion business model an unsustainable option in the future. The need to slow the pace of manufacture down is of huge importance for the industry with several high-designers recently acknowledging this. Sir Paul Smith for example has recently in the press expressed his concerns about the pace the industry is moving at; 'The world has gone mad. There's this absolute horrendous disease of greed and over-expansion and unnecessary, massive over-supply of product' (Barker 2016). Smith continues to discuss two key changes his brand will be making going forward; the paring down of the amount of collections offered across the brand changing to only two, and the amount of drops offered per year decreasing to four. Whilst this is still double the amount of the traditional fashion cycle of only two collections showing per year, it remains a step in a positive direction for the industry. Following the same strategy are Marc Jacobs and Burberry with nearly all brands and designers now constantly reinventing themselves in order to remain competitive. This reverse of strategies will see the industry moving back to the more traditional seasonal production, with spring/summer and autumn/winter collections being produced.

With this movement at the higher end of the fashion market, it has the potential to influence the lower, mass market end of the industry also. It is at the lower end however where the biggest positive changes can be made due to scale of production. The hope that this will filter into the fast fashion market is where real impact could be made, promoting a slower approach to fashion. This movement however would also require a change in mindset from consumers, who have been previously used to purchasing large quantities of cheaper clothing often. A move from wanting quantity to quality would be required with consumers purchasing fewer, higher quality products for longevity. This would naturally have price implications on garments which again would require a change in attitudes from consumers; however, fewer pieces which can be brought out season-after-season would have potential better price-per-wear qualities.

The consequences of this speed on the quality of design have also been acknowledged, with designers in industry being given as little as 25 min to come up with new collections. The implications of this time scale on the design process is irrevocable, with only copycat design work being achieved in this short timeframe (Rissanen 2016). The negative consequences on quality throughout the fashion process can be acknowledged, with a slower approach as suggested by Paul Smith having the potential to improve the innovation of fashion both in terms of design and production.

In addition to the fast fashion business model, it is the level of consumption of fashion, which also needs to be addressed. The continuous supply of new fashion items to high-street stores only encourages consumers to continue to purchase new products on even a weekly basis. This need to keep up with new, on-trend product is a reflection of the very nature of fashion, which continuously changes and at the high-end market level, should push boundaries in terms of innovation. This bi-seasonal approach has been abused, however, through the development of the fast fashion business model, turning the excitement of what the new season has to offer into a mundane drip-feed of average fashion goods. This change in the industry has only intensified the consumer need for more products continuously reinventing themselves and leading to a huge overconsumption issue.

The levels of consumption we refer to have not only huge negative implications on the sourcing and supply chain of the fashion process, but also causes issues at the post-consumer, end-of-life disposal of fashion. An estimated £100 million worth (based on 2015 prices) or around 350,000 tonnes of used clothing goes to landfill in the UK every year. In response to this, there are a large number of schemes being developed including that by WRAP who put forward the sustainable clothing action plan (SCAP), a 2020 commitment to encourage the fashion industry to be more eco-friendly. By 2013, 30 top high-street brands had registered which has now grown to 83 across a range or market sectors (WRAP 2015). The environmental impact of such large volumes of clothing being disposed of in landfill has huge negative implications, with many garments now being produced from non-biodegradable materials such as polyester. Due to the cheaper nature of fibres such as polyester, the fast fashion market level favours these fibres, which again tends to be the garments that are more readily disposed of. This disposable culture associated with the fast fashion market has caused the value of clothing to diminish with consumers having no attachment or tendency to care for such products. This loss of value on the part of the consumer again needs to be addressed and could be created through the slowing of the industry. Whilst it is not only price that can create value in clothing, if consumers were to pay a higher price for fashion items they purchased, their tendency to value and care for a product would increase. This would be again another reason that a more traditional two-season approach to the fashion industry would create a more sustainable future.

5   Recommendations for the Future

5.1   Addressing the Key Issues Preventing Change

There is a strong incentive for change in the fashion industry leading it to a more socially responsible and sustainable future. However as highlighted, there are several key factors that are currently preventing change which needs to be addressed in order to move things forward. Change in the industry will require cooperation from all stakeholders including retailers, consumers and manufacturers involved in the provision of the fashion product. Collaboratively, these parties can work together to change attitudes and approaches to fashion with the aim of slowing the industry down. As a consequence of this collaboration a shared vision of a more responsible future for fashion can be achieved.

When considering the issues preventing change, a number of recommendations can be made, yet it is to be acknowledged that these changes cannot be instant and that small developments over time will aid in this progression of the industry.

The first key issue discussed to be currently preventing change is the lack of consumer knowledge and awareness of ethical and sustainable issues in fashion. The absence of this knowledge not only has a negative impact on the choices that consumers make during the purchasing process but also means that the consumer has no comprehension of the negative implications of the manufacture of fashion products. This uninformed consumer can also not have any empathy or connectivity with the social element of the supply chain, or the understanding of the negative consequences this production has on the environment. In order to address this lack of consumer knowledge, fashion retailers need to begin to communicate their responsible intentions and actions. Despite the negative factors discussed with regards to the social and environmental compromise that occurs in the fashion supply chain, recent years have seen companies slowly changing their approach towards social responsibility. Many brands and retailers can now be seen to acknowledge these negative consequences and have began to implement goals and responsible actions as a result. As previously discussed, retailers have opportunities during the purchasing process to influence consumer behaviour for the better. The sharing of these goals and actions could help increase consumer knowledge and raise awareness of the steps the industry is taking to address the problems found. Increased communication could also help improve the consumer to retailer relationship and aid in development of this shared goal for a more responsible future. The improvement of this relationship has multiple benefits including the creation of brand trust and the power that consumer demand could have on moving the industry forward.

This increase of communication has began to slowly develop over the past few years with companies beginning to create communication channels such as annual reports to demonstrate their responsible intentions and actions. To help increase the effectiveness of these methods, the use of storytelling and narrative can be a powerful tool. The nature of this tool allows consumers to engage to a degree that suits them at the time, being able to dip in and out of the story to obtain pertinent details. It also helps to draw consumers in and aids engagement due to the structure of the text being used. This relates directly to another recommendation that can help in the increase of consumer knowledge and awareness, which is to ensure that when retailers do communicate with their customers regarding ethics and sustainability that they do so at the correct level. If the communication tool utilised uses heavy business language and jargon then a consumer will disengage with the message. Likewise if the method is too text heavy, consumers want to be visually interested and stimulated by the materials in order to successfully engage with the underlying messages.

The second key issue said to be preventing change is consumers not implementing their knowledge during the purchasing process. This relies on consumers who already have existing and pre-requisite knowledge of ethics and sustainability and the fact that they may use justification strategies to ignore what they know. Whether it be conscious or unconscious, the lack of connection between knowledge and the products being purchased is seen as a common issue when it comes to ethics and sustainability. Consumers often do not put the two together due to a lack of a relationship between products and their manufacturing supply chain. Therefore, it can often be overlooked to how a product has been made and the social and environmental consequences. In order to target this issue, fashion brands and retailers could increase the visibility of this relationship between a product and its responsibility. Commonly seen across the industry is the distinct separation between the products the company is retailing and their social responsibility goals and actions. Online, these are often positioned on a micro-site on their website provision taking customers away from their product purchasing site. This again indicates to the customer that these are two separate elements, failing to show the connection between them. In-store there is little communication to consumers other than retailers who provide separate ranges and collections, highlighting certain ethical and sustainable issues. As discussed previously, this can be seen as a negative approach to responsible business as again segregation away form mainline collections shows disparity between the responsible credentials. By increasing both the visibility in-store and online and emphasising the relationship between ethics and sustainability and the products, consumers will begin to create connections. This increase in connectivity should influence consumers during the purchasing process, making them think about the responsibility of a product prior to purchase.

The third key issue currently preventing the industry moving to a more responsible future is a lack of incentive for consumers in the engagement of responsible purchasing. As discussed earlier in this chapter, initiatives have been set up by some retailers in order to aid in addressing this. For example, Marks and Spencer carry out 'thank you' campaigns to encourage repeat good behaviour and engagement in responsible initiatives. In addition to these types of campaigns, retailers such as H&M and Marks and Spencer offer consumers vouchers of £5 to encourage them to bring old, unwanted clothes back to store for recycling purposes. Whilst money incentives will be effective, the underlying message behind these could be questioned due to the promotion of further consumption. Over and above these campaigns consumers have very little incentive to engage in responsible purchasing behaviour other than philanthropy. This relates directly back to a point previously made regarding Maslow's hierarchy of needs, where people move into to the final category of the pyramid. This is where people can begin to look past themselves and begin to seek out giving something back to society, which can be related to the premise of philanthropy. Relying on this as an incentive for responsible behaviour however is limiting, as only a small percentage of society will reach the final stage of Maslow's pyramid. Again in order to address this issue, the retailers are in a powerful position. Just as with the existing campaigns seen by Marks and Spencer and H&M, retailers need to be pro-active in their approach to engaging consumers in ethics and sustainability. Utilising social marketing techniques and engaging communication methods could aid in the promotion of these campaigns, whilst helping to inform consumers and raise their awareness. This again could have multiple benefits not only through increased engagement but also through increased knowledge promoting a more informed consumer, which in turn could influence more positive purchasing behaviour.

The fourth issue raised to be challenging the development of responsibility in the fashion industry is the segregation of certain collections or ranges which are more ethically or sustainably compliant than others offered in-store. As discussed earlier, whilst this approach has positive implications through the raising of awareness of certain issues, but it does send out mixed messages to consumers. The more switched-on consumers will begin to question this approach as to why only a small collection complies with certain responsibility standards and not others. This segregation could have very negative effects on a consumer's opinion and trust of a brand. In order to avoid this potential negativity, retailers need to implement an integrated business strategy in which they begin to build ethical and sustainable values into the way they do business. By incorporating this into their core underlying values, retailers will be able to offer their full product provision that satisfies certain responsible standards. This will prevent the segregation of products and promote brand trust. Consumers will no longer be presented with a choice and trade-off decision in preference to the choice of which retailers they choose to trust.

The adoption of an integrated business approach could also aid in addressing the lack of accessibility to socially responsible and sustainable products and also address the issue of the current provision being seen as not on-trend or fashionable. Through the elimination of two markets; responsible and non-responsible, the industry could begin to successfully move forward in addressing the key issues preventing change.

5.1.1 The Role of the Retailer

Throughout this chapter the role of the retailer has been heavily discussed, with reference made to their powerful position to influence change within the industry. As one of the three key stakeholders in the fashion process (manufacturers, retailers and consumers), they hold a unique position in which they have the ability to implement many strategies, techniques and approaches in the pursuit of a more sustainable future. However in order for this change to occur, brands and retailers need to utilise this powerful position and fully take advantage of the influence they can have over purchasing consumers and manufacturing producers. The relationship seen between the consumer and the retailer is varied and can change from company to company, however, in order to move things forward this relationship needs to be strong and be lead by the retailer. When successful this relationship has the power to heavily influence the products a consumer purchases, including the brand values they choose to buy into. Through the utilisation of social marketing techniques, retailers have the power to educate and inform the consumer about ethical and sustainable issues in fashion and help them to make informed decisions. This approach however needs to be handled with care and attention, as consumers do not want to feel they are subject to preaching regarding certain issues. This was highlighted by Catarina Midby, Head of Sustainable Communications for H&M; 'consumers go shopping to be inspired, not educated' (James 2015). This method of informing consumers also runs the risk of green washing, which is the use of marketing and PR to wrongly promote a companies products to being sustainable when they are not.

To avoid green washing as well as to ensure communication is effective, when retailers communicate with consumers they need to ensure that they pitch their information at the right level. This is to ensure engagement and interest in the consumer otherwise the communication is a pointless exercise. It is the role of the retailer here to know and understand their consumers, what they want from fashion and how much information they would like to know about the responsibility of the products they buy. The retailer needs to take responsibility for this and ensure that market research fully informs them of their target market. As well as the level the information is pitched at, the retailer needs to ensure that the methods of communication are correct for their target audience. Again through the detailed knowledge of their consumer's wants and needs, companies should be able to implement successful communication techniques.

In addition to the informant, the retailer also needs to adopt the role of providing a sound rationale for the engagement in ethics and sustainability. As seen with many companies, it is the responsibility of the retailer to not only provide engaging campaigns and initiatives for the consumer to engage in but also to provide incentives to do so. Philanthropy alone is not enough for successful engagement and it is the job of the retailer to ensure that again their incentives are appealing and pitched at the correct level for their target market. If successful, the retailer could not only get their consumers to engage but they could further inform them of ethical and sustainable issues in the process, providing multiple benefits to this approach being utilised.

Above all however it is the role of the retailer to successfully engage in ethical and sustainable pursuits, whilst remaining profitable and desirable to the market. The Global Ethical Trading Manager for Monsoon and Accessorize reiterated this point; 'whilst profit might be a dirty word, if we do not have a commercially successful business then we cannot put pressure on our suppliers to make changes and its trying to create that necessary balance and harmony' (James 2015).

6 Conclusion

6.1 Summary

As this chapter has demonstrated the fashion supply chain is a long and complex process that does not come without its issues, the complexities of this process leave it wide open to both environmental and social compromise adopting many forms. Rana plaza is just one of the many social disasters to have occurred in recent years, with significant consequences to the many humans who engage in the production of fashion. However the social compromise is just half the story, with the supply chain rapidly eating away at natural resources such as water and oil in the production and maintenance of fashion textiles. Evidence such as that provided earlier in the chapter regarding the Aral Sea demonstrates the vast water usage that the production of natural fibre cotton garments requires. The manufacture of a single pair of denim jeans accounts for 11,000 L of water in the growth of the cotton fibre, garment processing and dying of the denim fabric itself (WWF 2015). This statistic of water usage however does not account for the maintenance and washing of his product, which can be seen as the most accountable stage in the fashion life cycle.

The need for change has never been more prominent. As the fast fashion business model continues to flourish, the time pressures on the fashion supply chain only intensify. For example, when the Rana Plaza disaster occurred in April 2013 it was thought that it would have a negative impact on sales of fast fashion goods due to extensive media coverage. This however was quite the opposite, with Primark boasting a 44 % increase in like-for-like sales on the previous year (Hawkes 2013). Social compromise in the supply chain along with the vast consumption issues in social fashion cannot continue with urgent developments needing to be implemented to bring about this change.

With the fashion production cycle there are three key stakeholders; the manufacturers, the retailers and the consumers. In the context of the purchasing process, both the consumer and the retailer engage and interact with each other through a mutual relationship. Despite this relationship being two-way to a certain degree, the retailer ultimately has the power to have a degree of influence over the products being purchased by the consumers. Traditionally this influence relates to maximising profitability and increasing sales through the utilisation of marketing techniques. However when recontextualised this power has the potential to influence purchasing behaviour for the good. As demonstrated in Fig. 3, this influence is most effective during the designated window of opportunity. This lies between the consumer having formulated their initial intent and this initial intent translating into behaviour. However as previously identified, it is also at this stage where a fundamental difference between intentions and behaviour can occur. The intention-behaviour gap, as shown in Fig. 6, is where a consumer has the intention to purchase responsibly but where in this intervening period their decision is changed and consequently their behaviour is different from that initially intended. When developing this intent, first the need recognition has to be identified; this is followed by the consumer conducting an initial scoping exercise where products that will fulfill this need are to be found. The final stage in developing a purchasing intention is the examination of competition, trying to find the best product for the best price. It is at this stage that the retailer has the power to intervene and influence purchasing behaviour before the final product/s are bought. When considering social responsibility and sustainability, this stage in the purchasing process allows for retailer intervention to utilise more social marketing tools, where behavior could be influenced for the good. Executed correctly, this could help the industry begin to move forward to a more socially responsible future through consumers making more informed choices in the products they choose to buy.

In order for this influence of behaviour to be effective, the relationship between the retailer and consumer is vital for success. Communication needs to be carefully considered on the part of the retailer in order to determine the best methods to be utilised in influencing consumer behaviour for the good. As previously discussed, in the past retailers have been accused of green washing in their approach to marketing, which falsely sees companies using green marketing to promote products responsible criteria. The balance of informing and preaching is also an issue, which retailers have to be wary of, with this approach also said to be putting consumers off. The careful utilisation of social marketing methods could aid this relationship and help inform consumers of ethical and sustainable issues in fashion, leading to more informed decisions being able to be made. Again however in order for this approach to be successful, the retailer must be seen in being pro-active in their communication and marketing actions. This engagement of the consumer is again a vital factor for moving the industry forward, as all stakeholders in the fashion process must buy into the values, in a shared vision approach.

The retailer is in a unique and powerful position and must be the driving force behind this change and encourage both consumers and manufacturers to engage in their responsible values. The role of the retailer is not only to inspire but also to inform and engage their customers in their goals for a more socially responsible fashion industry moving forward.

Appendix

Consumer Purchasing Hierarchy

During the analysis process, patterns were identified where it became apparent there was a top/bottom divide between the eight choices provided. The top four choices, aesthetics, material, price and washing instructions, could be described as necessities when purchasing clothing, in comparison to the bottom 4 choices; handmade, locally sourced, Fairtrade and organic, which could be described as desirable factors, however, not necessary when deciding to purchase an item.

The following results have been divided by the number of choice to provide greater details and insights.

First Choice

13 of the 15 participants voted *aesthetics* as their priority choice when purchasing garments, showing that the look and image of garments was the most important factor to the participants when buying fashion items. Whilst one participant detailed their first choice as *material,* the final participant chose *locally sourced*; however, this individual did differ in personal details in comparison to the majority of the group. This answer could therefore be classified as an irregularity (Table 1).

Second Choice

The second choice for a large proportion of participants was *material* with eight out of 16 people voting this way. From the qualitative reasoning provided, participants saw material as very important due to it being an indication of quality. There were also two votes for *organic*, whilst again these participants did differ in age and salary range to the large part of participants. The remainder of votes was dedicated to *price* (Table 2).

Table 1 Garment requirements—1st choice results

Choice factors	Aesthetics	Price	Material	Hand made	Locally sourced	Fairtrade	Organic	Washing instructions
Participant votes	13	0	1	0	1	0	0	0

Table 2 Garment requirements—2nd choice results

Choice factors	Aesthetics	Price	Material	Hand made	Locally sourced	Fairtrade	Organic	Washing instructions
Participant votes	1	5	8	0	0	0	2	0

Third Choice

The third choice shared the votes between *material* and *price*; however at this point of the participant, hierarchical choice was the first votes for *Fairtrade* and *washing instructions*. This indicates that by the third choice, consumers are beginning to incorporate ethical attributes into their purchasing rationale. By this third hierarchical choice, all participants had voted for *aesthetics,* meaning that it was in the top three priority choices by all the participants in the activity (Table 3).

Fourth Choice

8 of the 16 participants chose *washing instructions* as their fourth priority choice in the garment requirements activity. This category also saw the first votes for *handmade* which provided evidence for the top/bottom divide beginning to emerge. This choice also saw the final vote for *materials,* which meant that this was in the top four choices for all participants (Table 4).

Fifth Choice

The final choice saw *Fairtrade* and *handmade* as the most popular choices, illustrating that people saw these factors as desirable but not vital when considering their purchasing criteria. As shown in Table 5, the votes have shifted to the choices

Table 3 Garment requirements—3rd choice results

Choice factors	Aesthetics	Price	Material	Hand made	Locally sourced	Fairtrade	Organic	Washing instructions
Participant votes	1	7	6	0	0	1	0	1

Table 4 Garment requirements—4th choice results

Choice factors	Aesthetics	Price	Material	Hand made	Locally sourced	Fairtrade	Organic	Washing instructions
Participant votes	0	1	1	1	2	2	1	8

Table 5 Garment requirements—5th choice results

Choice factors	Aesthetics	Price	Material	Hand made	Locally sourced	Fairtrade	Organic	Washing instructions
Participant votes	0	2	0	3	2	4	2	2

to the right hand side of the table, where the ethical and sustainable attributes were more apparent. When compared the table of results in first choice, the majority of votes was positioned at the left-hand side of the table, indicating again this divide between necessity and desirable factors when purchasing garments.

References

Ajzen, I. (1985). *From intentions to actions: A theory of planned behaviour.* New York: Springer-Verlag.

Arnold, C. (2009). *Ethical marketing and the new consumer.* West Sussex: Wiley.

Auger, P., & Devinney, T. M. (2007). Do what consumers say matter? The management of preferences with unconstrained ethical intentions. *Journal of Business Ethics, 76,* 361–383.

Barker, A. (2016). *Streamlining collections, paul smith reveals own fashion calendar fix.* Available at: http://www.businessoffashion.com/articles/intelligence/streamlining-collections-paul-smith-reveals-own-fashion-calendar-fix. Accessed 8 Feb 2016.

Barrie, L. (2009). *Just style.* Available at: www.just-style.com/comment/spotlight-onsustainability-in-design_id105739.aspx. Accessed 11th Jan 2011.

BBC. (2013) 'Bangladesh building collapse death toll over 800', BBC News, 8 May 2013 [Online]. Available at: http://www.bbc.co.uk/news/world-asia-22450419. Accessed 8 May 2013.

Belk, R., Devinney, T. M., & Eckhardt, G. (2005). Consumer ethics across cutures. *Consumption, Markets and Culture, 8,* 275–289.

Bhuiyan, K. (2012) 'Bangladesh Garment Factory Disaster Timeline', Compliance – Updates, 29 November 2013 [Online]. Available at: www.Steinandpartners.com/sustainability/compliance/bangladesh-garment-factorydisaster-timeline. Accessed 30 January 2014.

Bray, J., Johns, N., & Kilburn, D. (2010). An exploratory study into the factors impeding ethical consumption. *Journal of Business Ethics, 98,* 597–608.

Butler, S. (2013) 'Bangladeshi factory deaths spark action among high-street clothing chains', The Guardian, 23 June 2013 [Online]. Available at: http://www.theguardian.com/world/2013/jun/23/rana-plaza-factory-disaster-bangladesh-primark. Accessed 26 August 2013.

Carrigan, M., & Attalla, A. (2001). The myth of the ethical consumer–do ethics matter in purchase behaviour? *Journal of consumer marketing, 18,* 560–578.

Carrington, M., Neville, B., & Whitwell, G. (2010). Why ethical consumers dont walk their talk: Towards a framework for understanding the gap between the ethical purchase intentions and actual buying behaviour of the ethically minded consumers. *Journal of Business Ethics,* 139–158.

Chatzidakis, A., Hibbert, S., & Smith, A. P. (2007). Why people dont take their concerns about fair trade to the supermarket: The role of neutralisation. *Journal of Business Ethics,* 89–100.

Clouder, S., & Harrison, R. (2005). The effectiveness of ethical consumer behaviour. In R. Harrison, T. Newholm, & D. Shaw (Eds.), *The ethical consumer* (pp. 89–106). London: Sage Publications.

Cowe, R., & Williams, S. (2001). *Who are the ethical consumers?* Manchester: Co-operative Bank.

Davies, C. (2007). Branding the ethical consumer. *The Financial Times,* p. 18, 21 Feb.

Devinney, T. M., Auger, P., & Eckhardt, G. (2010). *The myth of the ethical consumer.* Cambridge: Cambridge University Press.

Devnath, A., & Srivastava, M. (2013) "Suddenly the Floor Wasn't There,' Factory Survivor Says', Bloomberg, 25 April 2013 [Online]. Available at: http://www.bloomberg.com/news/2013-04-25/-suddenly-the-floor-wasn-t-therefactory-survivor-says.html. Accessed: 28 April 2013.

Dickson, M. A. (2013). Identifying and understanding ethical consumer behavior: Reflections on 15 years of research. In: J. Bair, M. Dickson & D. Miller (Eds.), *Workers' rights and labor compliance in global supply chains* (pp. 121–139). Routledge: New York.

Ellen, P. (1994). Do we know what we need to know? Objective and subjective knowledge effects on pro-ecological behaviours. *Journal of Business Research, 30*(1), 43–52.

Fletcher, K. (2008). *Sustainable fashion and textiles.* London: Earthscan.

Genecon LLP and Partners. (2011). Understanding high street performance, 5 Dec (Online). Available at: www.gov.uk/government/uploads/system/uploads/attachment_data/file/31823/11-1402-understanding-high-street-performance.pdf. Accessed 12 Apr 2012.

Giesen, B. (2008). *Ethical clothing: New awareness or fading fashion trend?*Germany: VDM Publishing.

Hawkes, S. (2013). People thought Rana Plaza would be a blow to primark. Today's profit figures say otherwise. *The Telegraph*, 5 Nov 2013 (Online). Available at: http://blogs.telegraph.co.uk/news/stevehawkes/100244468/people-thought-rana-plaza-would-be-a-blow-to-primark-todays-profit-figures-say-otherwise/. Accessed 30 Jan 2013.

Hawkins, D. E. (2006). *Corporate social responsibility*. Hampshire: Palgrave MacMillan.

James, A. M. (2015). *Influencing ethical fashion consumer behaviour—A study of uk fashion retailers.* Unpublished Ph.D. thesis. University of Northumbria at Newcastle.

Kotler, P., & Lee, N. (2005). *Corporate social responsibility. Doing the most good for your company and your cause*. New Jersey: Wiley.

Lee, M. (2007). *Eco chic*. London: Octopus Publishing Group.

Marks and Spencer. (2015). *Shwopping*. Available at: http://corporate.marksandspencer.com/plan-a/our-stories/about-our-initiatives/shwopping. Accessed 1 Mar 2016.

Mintel. (2007). *Green & ethical consumer*. Mintel: London.

Mintel. (2009). *Ethical clothing*. Available at: http://academic.mintel.com. Accessed 27th Nov 2010.

Morgan, L., & Birtwhistle, G. (2009) An investigation of young consumers disposal habits. *International Journal of Consumer Studies*, 190–198.

Nelson, D., & Bergman, D. (2013) 'Scores die as factory for clothing stores collapses', The Independent 25 April 2013 [Online]. Available at: http://www.independent.ie/world-news/asia-pacific/scores-die-as-factory-for-clothingstores-collapses-29220894.html. Accessed 25 April 2013.

Newholm, T., & Shaw, D. (2007). Studying the ethical consumer: A review of research. *Journal of Consumer Behaviour, 6*, 253–270.

Niinimaki, K. (2010). Eco-clothing, consumer identity and ideology. *Sustainable Development, 18*, 150–162.

Ozcaglar-Toulouse, N., Shiu, E., & Shaw, D. (2006). In search of fair trade: Ethical consumer decision making in france. *International Journal of Consumer Studies, 30*, 502–514.

Portas, M. (2011). The Portas review: An independent review into the future of our high streets, (Online). Available at: http://www.bis.gov.uk/assets/biscore/business-sectors/docs/p/11-1434-portas-review-future-of-high-streets.pdf. Accessed 4th Feb 2012.

Rest, J. R. (1986) *Moral development: Advances in research and theory*, New York: Praeger.

Rissanen, T. (2016). Slow down! We are creative. In *Creative Cut Conference*, University of Huddersfield, 24 Feb.

Ritch, E., & Schroder, M. (2009). *What's in fashion? Ethics? An exploration of ethical fashion consumption as part of modern family life*, (Online). Available at: http://www.northumbria.ac.uk/static/5007/despdf/events/era.pdf. Accessed 24th Sept 2010.

Schiffman, L., Kanuk, L., & Hansen, H. (2008). *Consumer behaviour—A European outlook*. Essex: Pearson Education.

Soloman, M., & Rabolt, N. (2004). *Consumer behaviour in fashion*. New Jersey: Pearson Education.

Szmigin, I., Carrigan, M., & Mceachern, G. (2009). The conscious consumer: Taking a flexible approach to ethical behaviour. *International Journal of Consumer Studies*, 224–231.

Tokatli, N. (2007). Global sourcing: Insights from the global clothing industry—The case of Zara, a fast fashion retailer. *Journal of Economic Geography,* (Online). Available at: http://www. nihul.biu.ac.il/_Uploads/dbsAttachedFiles/zara06.pdf. Accessed 30th Sept 2010.

WGSN. (2010). Final edit 2010. *World Global Sourcing Network,* 28 Sept, (Online). Available at: www.wgsn.com. Accessed 29 Nov 2011.

Worcester, R., & Dawkins, J. (2005). Surveying ethical & environmental attitudes. In R. Harrison, T. Newholm, & D. Shaw (Eds.), *The ethical consumer* (pp. 189–203). London: Sage Publications.

WRAP. (2015). SCAP 2020 commitment. *WRAP* (online). Available at: http://www.wrap.org.uk/content/scap-2020-commitment. Accessed 14 Mar 2016.

WWF. (2015). The hidden cost of water. *WWF* (online). Available at: http://www.wwf.org.uk/what_we_do/rivers_and_lakes/the_hidden_cost_of_water.cfm. Accessed 1 Mar 2016.

To Fur or not to Fur: Sustainable Production and Consumption Within Animal-Based Luxury and Fashion Products

Mukta Ramchandani and Ivan Coste-Maniere

Abstract We live in the age of information technology where information travels faster than the speed of light. When a slightly inclined sustainable consumer searches for ethical fashion and luxury brands, they are easily bombarded with advertisements and information. The ongoing trends in adopting sustainable consumer lifestyles, being green and ethical, add to the lustre of modern-day consumers. But despite the awareness, an alarming increase of 70 % in the global sales of the fur industry in the past decade has contradicted the sustainable luxury and fashion movement. Where on the one hand, 100 % fur-free fashion companies like Stella McCartney, Tommy Hilfiger, Calvin Klien and Ralph Lauren are setting an example in the fashion industry. But on the other hand, companies like Gucci, Donna Karan and Karl Lagerfeld have made fur as their forefront in the fashion shows. Interestingly, in the luxury industry the big quest has been about understanding if sustainability and luxury can co-exist and how sustainability can be defined in the realms of luxury? But it is difficult for consumers to adhere to a reference point when it comes to using sustainable animal-based products. In general, the supply chain and fair trade has been an important aspect of eco-fashion products but how does it fit for animal-based products like fur is not well understood. The luxury and fashion industry caters to both sustainable and

M. Ramchandani (✉)
Neoma Business School, Reims, France
e-mail: muktaramchandani@gmail.com

M. Ramchandani
Olten, Switzerland

I. Coste-Maniere
Luxury and Fashion Management, SKEMA Business School, Sophia Antipolis, France
e-mail: ivan.costemaniere@skema.edu

I. Coste-Maniere
Luxury and Fashion Management, SKEMA Business School, Suzhou, China

I. Coste-Maniere
Global Luxury Management, SKEMA Business School, Raleigh, USA

I. Coste-Maniere
Luxury Retail in LATAM, Florida International University, Miami, USA

© Springer Science+Business Media Singapore 2017
S.S. Muthu (ed.), *Textiles and Clothing Sustainability*,
Textile Science and Clothing Technology, DOI 10.1007/978-981-10-2131-2_2

non-sustainable consumptions. In this chapter, we unfold the realities of the fur and faux fur industry. We examine what has led to the come back of fur within the age of sustainable luxury and fashion through interviews from the industry experts and secondary literature. Developing on the industry data and interviews we show the technicalities from production and consumption cycle of the fur industry. We explore the consumer profiles into the consumption of fur and faux fur products. We elucidate how men and women differ within these consumption patterns.

Keywords Fur · Faux fur · Eco-fashion · Sustainable luxury and fashion · Animal welfare

1 Introduction

Throughout centuries, fur pelts from animals like mink, fox, cats, dogs, bears, racoons, etc., have been worn for their warmth and traded across the world. According to the Russian fur history (Sojuzpushnina 2016), in Russia, fur served as a form of currency, it was used as gifts and as part of a bride's dowry, and became a significant part of trade during the tenth and eleventh centuries. In the 1530s, the beaver became a main trading item between the American Indians and the colonists, and beaver pelts were regularly shipped to Europe. By the late 1500s, fur was extremely popular in Europe (Peterson 2010). In 1608, Samuel de Champlain, a French explorer, created a trading post in Quebec, which became the centre of fur trade in America. In the seventeenth century, Siberia's unification with Russia helped to propel Russia to become the largest fur supplier, which it remained until the nineteenth century. Around that time, fur farming started in North America, and was introduced into Europe in the early twentieth century (Peterson 2010). According to British Fur Trade Association, the global fur trade has rose to 58 % since the 1990s. The global fur trade is estimated to be more than $40 billion (International Fur Federation 2016). Fur farming is valued at $7.8 billion and total employment in the sector at over one million (International Fur Federation 2016).

The softness and durability have been improved nowadays with advanced processing and dyeing techniques, which pro-fur argumentators advocate as being sustainable compared with the faux fur. However, formaldehyde, ammonia, hydrogen peroxide and other bleaching agents are used to dye furs which are found to be very dangerous for the environment. Countries like China which is one of the largest producers of fur do not have effective laws against the harsh chemicals used (antifurcoalition.org).

2 Fur as a Luxury

The reason fur is considered a luxury is due to the fact that fur is handcrafted, requiring the skills of trained artisans who understand the qualities of fur and the special techniques that go into creating patterns, blocking and sewing fur. The craftsmanship defines the timelessness and exclusivity of the product which is why fur is designed to last for longer time periods. In the realms of sustainability, does that make fur a sustainable product?

This chapter answers how the animal-based products flourish within the sustainable luxury and fashion industry. To maintain the ethicality of sustainable brands what considerations do companies have? And how consumers are motivated to make purchase decisions for such kind of products? From theoretical standpoints how does status motivate the consumptions? Where is the right balance between production and consumption? We aim to answer and elaborate on these questions and the underpinning aspects of the fur industry.

2.1 Types of Natural Fur

Following sections describe various types of natural fur.

2.1.1 Wild Fur

Wild fur is less expensive than farmed fur due to the less control on the damages like scratches caused on the fur from its natural wild environment. Indeed, some consumers prefer to wear wild fur knowing that it comes from natural habitat of animals than animals raised in a cage.

2.1.2 Invasive Fur

In the coastal areas of the US, nutria a rodent has been declared as an invasive animal (Avnis 2015). Since the 1930s nutrias, originally from South America, have been gobbling up the wetlands of coastal Louisiana, contributing to land loss that approaches 25 square miles per year, along with billions of dollars (Avnis 2015). Since 1990s the Louisiana Department of Wildlife and Fisheries created an incentive programme that they would pay registered hunters and trappers four dollars for each nutria they killed. With a fashion project called Righteous Fur, Cree McCree, a New Orleans based writer and artist is using fur from nutria to better use the dead animal rather than just letting them being thrown into the swamps.

2.1.3 Farmed Fur

Fur farming by some is considered to be a sustainable practice as it involves the fur farmed animals to be fed food wastes from humans. This helps in keeping down the costs of food production and helps in reducing wastage (Fur Institute of Canada). Fur farms make productive use of discarded lands. Raising fur animals suits well for mixed farming since during the winter months the demand for field crops is not so high and needs less attention from the farmer. To insulate cages and make beddings straws from crops are used, while the manure from ranched animals is used for the soil as fertilizer. Fur farmers also use farm wastage as a source of bioenergy to power their own farms. As a renewable natural resource and recycler, farmed fur is considered a sustainable product (Fur Institute of Canada). But fur farming is banned in countries like Austria, Switzerland, United Kingdom, the Netherland, Slovenia, etc.

2.2 Fashion and Fur

In 2015 at Haute Couture events and fashion weeks the trend in the fur fashion has been with vintage design and retro style silhouettes. Due to the global warming short fur coats in jackets are as well becoming a trend like fur boleros, furry vests and short sleeves cropped jackets. Fashion houses like Stella McCartney, Zara, Hugo Boss and H&M are fur free. Recently, in 2016 Georgio Armani declared to stop using fur, who was hailed by the anti-fur brigade but criticised by fellow pro-fur fashion designers like Karl Lagerfeld. Fashion houses like Alexander McQueen, Dolce and Gabbana, Karl Lagerfeld, Michael Kors, Fendi, Oscar de la Renta, Prada, Vivienne Westwood and Yves Saint Laurent very openly promote the use of fur through their designs. New designers like the Lithuanian-based designer Josef Statkus who was awarded twice by LVMH for his haute couture designs do not hesitate to use fur. In fact, new designers are supplied fur from the fur federations to be able to use in their future designs.

2.3 Celebrities and Fur

While in the age of social media some celebrities endorsing fur are proudly welcomed, some are even condemned and backlashed. Like Brazilian celebrity Cristina Cordula famous for her French tv show "les reines du shopping" angered her fans when during an interview she openly said that she is not against the use of fur and the suffering of farmed fur animals is another debate. Her fans were shocked and expressed their anger through social media.

3 Methodology

For our research, we have used qualitative research methodology. Secondary data also includes literature from various research journals and sources of information covering different points of view on trends of the industry like newspaper, journal and magazine articles. This also included several web sources such as blogs, online articles and company web sites. In addition to this collation and parsing of information, we have drawn upon our discussions with various public domains for reference. The expertises of the contributors' long-term experience and existing relationships in the industry have been very useful. Interviews are one of the key methods to collect up-to-date information from the experts in the industry. We interviewed fur industry expert and businessman from Russia giving us insights of prevalent questions for our research, as well as scientific reports from the fur industry have been utilised. As a result, we looked at the production cycle of fur, focusing on natural and faux fur. Taking some examples and depending on their engagement in sustainability, we explored some of the remarkable works being done already. Finally, we analysed the outcome seeking to identify whether sustainability is a challenge or an opportunity for the fur sector.

4 Theoretical Approach to Research

4.1 Derivatives of Sustainable Behaviour

From a theoretical perspective, in this section we discuss on what leads consumers to consume sustainable products which will help the readers understand it in the context of the fur industry.

4.1.1 Costly Signals

Costly signalling theory explains that (Miller 2000; Zahavi 1975) both animals and humans often engage in altruistic acts, acts that seem to involve a sacrifice and primarily to convey or communicate a signal about themselves (Gintis et al. 2007). For example, individuals often enact some altruistic behaviour to show they are elevated in status, called competitive altruism.

4.1.2 Need for Status

Scholars have defined an individual's susceptibility to interpersonal influence as "the need to identify with, or enhance one's image in the opinion of significant others through the acquisition and use of products and brands, and the willingness to conform to the expectations of others regarding purchase decisions" (Bearden et al. 1989, p. 473). Studies have found that status-seeking consumers are concerned with their peers and use brands to convey status (Ruvio et al. 2008). Griskevicius et al. (2010) discussed how people indulge in sustainable luxury just for the sake of status; a desire for status can spur self-sacrifice and present a powerful tool for motivating prosocial and pro-environmental action.

4.1.3 Need for Uniqueness

The theory of consumers' need for uniqueness explains how an individual's need for uniqueness can influence brand responses and the need to be different from others (Ryan 2008; Tian et al. 2001) through the pursuit of material goods (Knight and Kim 2007). For example, when haute couture brought back fur as a huge success since 2014 and 2015 at various fashion shows in US and Europe, consumers from China followed these fashion trends, which lead to a high demand of fur. Beyond the trend and status symbol, fur is giving people need for uniqueness and plays an important part of their self-esteem and identity.

4.1.4 Anthropocentricism

Values are preceded by culture, society and personality, and have behaviour as its outcome (Rokeach 1973). Anthropocentric people show environmental concern due to the reason that they think human comfort, quality of life and health can be dependent on the preservation of the environment (Gagnon Thompson and Barton 1994). People are notoriously reluctant to change familiar patterns of behaviour, and making a switch to green behaviours often necessitates making sacrifices (e.g., paying more for a less-effective product) (Griskevicius et al. 2010). Young generation with being more educated and informed is avoiding the use of fur as they are becoming more anthropocentric.

5 Interviews

For our research on the fur industry we interviewed fur industry expert and businessman Farit Mullayanov from Moscow, Russia. Following are the questions and responses from the interview:

1. *Where are the major consumers of fur based? In Which countries?*
   The main consumers of the fur garments live in the cold climate conditions. As in the areas, where it is cold, one simply cannot survive in fact without real fur. In the northern areas of Russia, like Arkhangelsk, Murmansk, Khanty-Mansiysk, Koryak, Chukotka, Yakut regions or Siberia, fur clothes are considered to be even

the tradition and are used in the everyday life (not only winter) already for centuries. When it comes to the foreign countries we should think northern regions of Norway, Iceland and Canada where the pieces are used.

2. *What are the consumer profiles indulging in consumption of fur? Are there any certain age group indulging more?*
   Young people mostly buy garments made of sheep fur (mutton) as it is a cheaper option. Older people can afford and buy more luxurious clothes made of mink or sable.

3. *More consumers of fur are men or women?*
   Main consumer group is women. Men use less pieces of fur garments.

4. *Is sustainable faux fur a threat to real fur?*
   Faux fur does not replace the real fur as it does not possess the warming ability of the real fur and does not serve its main purpose.

5. *What kind of people will switch from real fur to faux fur?*
   Those who live in the regions where it is possible to survive without the real fur can replace it with the faux fur. Citizens of the regions from the northern Russia and Siberia are not able to switch to the faux fur because it will not help them with what they use the real fur for.

6. *Why is there an increasing trend in the fur industry despite the increase in awareness from PETA and other animal rights organisations?*
   Currently, there is no trend on any increase in the consumption in the fur industry. Quite the contrary, the industry suffers the low and lowering demand. The prices of the raw materials are constantly decreasing as a result as well.

7. *In your opinion what will be the future trends in the fur industry?*
   In the coming years the fur industry will be declining its production capacity. There are two main reasons for that:

– global warming which is very highly seen in the recent years
– invention and increasing production of the warmth-producing materials (not faux fur) which still perfectly serve the function to provide heat and are a compatible substitute to the real fur.

The data obtained from Mr. Mullayanov takes our research into deeper understanding of the fur industry's consumption patterns. The next sections will help readers understand different facets of production.

6 Production Cycle in the Fur and Leather Industry

6.1 How Fur Is Produced?

Some of the techniques used by furriers and manufacturers to lighten the garment are shearing, plucking, knitting, leathering and weaving. The most important centres for fur manufacturing include Canada, the Chinese mainland, Greece, Hong Kong and Russia, Germany, Italy, Korea, Japan, Spain, Turkey, Ukraine and the US.

6.1.1 Steps in Production of Fur

(1) Killing/Slaughtering of animal
    In the EU, Council Directive 98/58 sets down rules covering the welfare of all
    farmed animals, including fur farmed animals, while Regulation
    (EC) No. 1099/2009 deals with the slaughter and killing of farmed animals
    including fur animals (HKTDC 2016). Electrocution and gas inhalation
    method are a commonly used method for killing fur animals. Carbon
    monoxide is recommended usually but sometimes it can be slow to induce
    effects of unconsciousness in minks. The time taken to induce unconscious-
    ness in minks is 64 s for CO ($\geq$7 %) and 76 s for 100 % N_2 as opposed to 19 s
    for 100 % CO_2 and 26 s for 70 % (Hansen et al. 1991). Other methods include
    injections of chloral hydrate solution and breaking the neck.
(2) Skinning
    After the animal is slaughtered the skinning takes place and pelts are prepared
    for auction. Skinning can be an automated process. After skinning the pelts are
    fleshed and placed inside out on a board for stretching and drying which takes
    about 3–4 days (Bijleveld et al. 2011).
(3) Auction houses
    Majority of fur farms are found in Denmark, China, Netherlands, Baltic States
    and the USA. Commonly, raw skins produced by fur farmers and trappers are
    sold through modern international auction houses, often located close to
    producing areas (International Fur Federation 2016). The world's largest fur
    auction houses are in Copenhagen, Helsinki, St. Petersburg, Seattle and
    Toronto. Trade fairs like International Fur and Leather Exhibition (MIFUR) in
    Milan and HKIFFF in Hong Kong have been very successful recently for the
    splurge in the fur industry. Kopenhagen fur which constitutes 60 % of the
    global market share (Orange 2014) sold mink skins for up to $2.4 billion in
    2013.
(4) Further processing:
    Due to the preservation techniques used, raw pelt is dry and hard. After
    auctioning the raw fur is converted into leather and renders it useful for
    garments. To get the desired look the leather may then be dyed with bleaching
    agents. It is similar to leather production except that the animal hairs are
    conserved (BASF 2010).

6.1.2 Apparel Manufacture

In Europe, important fur apparel manufacturing locations are Kastoria and Siatista
and the surrounding area, in Greece (Bijlevel et al. 2011). The steps of apparel
manufacture are as follows (Connecticut Furs Inc. 2016):

- selection of the number of furs needed for the desired design;
- slicing the skin into strips and sewing these together to make the designed pattern;
- soaking in water, stretching and drying, to match the form and design of the pattern;
- mounting additional parts, like closures.

6.1.3 Faux Fur Production

Faux fur fibre is produced from petro-chemicals as part of large integrated chemical manufacturing facilities. Europe, Japan and North America account for much of the annual global production (DSS Management Consultants Inc. 2012). Considerable processing is required to convert acrylic fibre into faux fur fabric. Further, the actual production of faux fur fabric often occurs quite distant from where the fibre is produced, example China is a major faux fur fabric producer.

7 Environmental Impact of Natural Versus Faux Fur Coat Production

In this section we discuss various environmental impacts of natural and faux fur and describe the literature associated with the analysis of the environmental impacts.

7.1 Life Cycle Assessment

According to US Environmental Protection Agency (USEPA 2016), Life Cycle Assessment (LCA) is a technique to analyse the environmental impacts of a product's life from material creation to disposal or recycling. The systematic approach of LCA is utilised by various industries and academicians (Willaims 2009). LCA aids analysts (Williams 2009) in the following purposes:

- calculate a product's environmental impact,
- identify negative and positive environmental impact of processing or production,
- find opportunities for improvement,
- compare and analyse several processes depending on their environmental impacts,
- quantitatively justify a change in process or product.

There are four main phases of LCA which are defining goals and scope, inventory analysis, impact assessment and interpretation (ISO 14040 2006 and ISO 14044 2006). A large number of LCA literature describe various indicators (Gyetvai 2012); these are

(1) Selected life cycle impact (LCI) indicators—This is useful to track quantity flows like the use of secondary energy throughout product's life cycle. These are not impact indicators directly but are useful for interpretation phase of a LCA study.
(2) Midpoint life cycle impact assessment (LCIA) indicators—The midpoint LCIA indicators (or potential indicators) characterise various environmental problems like climate change, ozone depletion, photochemical ozone formation, acidification, eutrophication and resource depletion.
(3) Endpoint LCIA indicators—Endpoint LCIA indicators refer to actual damage categories like damage to resources, damage to human health and damage to the ecosystem.

Examples:
Bijleveld et al. (2011) from CE Delft conducted a report on the environmental impact of mink fur production. According to this report, 11 animals are required to produce 1 kg of fur. In the course of its lifetime, mink eats about 50 kg of feed, resulting in 563 kg of feed required per kg of fur. Although the feed consists mainly of offal and this is accounted for by very low allocation of environmental impacts, the 563 kg required to produce 1 kg of fur knocks on considerably in the total environmental footprint of fur and for 14 of the 18 impact categories studied, feed is the predominant factor. Compared with textiles, fur has a higher impact per kg in 17 of the 18 environmental categories, including climate change, eutrophication and toxic emissions. In many cases, fur has impacts that are a factor 2–28 higher than textiles, even when lower bound values are taken for various links in the production chain.

According to the LCA report by DSS Management Inc. (2012) faux fur coat production has higher environmental demands than a natural fur coat production. As illustrated in Table 1, a comparative analysis on environmental impact by

Table 1 Life cycle scores and percent differences for individual midpoint indicators (*Source* DSS Management Consultants Inc. 2012)

Impact category	Natural Fur raw score	Fake Fur raw score	Percent difference (%)
Carcinogens	4.096	7.960	94
Non-carcinogens	3.92	5.200	32
Respiratory inorganics	86.971	84.131	−3
Ionising radiation	0.246	1.159	370
Ozone layer depletion	0.065	0.040	−39

Note All scores are reported in 'millipoints' units. Millipoints is an abstract unit used to express diverse types of potential impacts. Refer to the Impact 2002+ website for further details. University of Michigan Risk Science Center—Risk and Impact Modeling—Research—Impact 2002+

natural fur coat and fake fur coat shows that a faux fur coat scores significantly better for three indicators, namely, respiratory organics emissions, ozone layer depletion and terrestrial acidification/nitrification. On the other hand, the life cycle of a faux fur coat results in considerably greater consumption of non-renewable energy, greater risk of potential impacts of global warming and greater risk of potential impacts from ionising radiation. As well, there is greater risk of potential impacts from carcinogenic and non-carcinogenic emissions and greater risk of potential terrestrial ecotoxicity impacts with the life cycle of a faux fur coat.

Comparative analysis of environmental impact between natural fur and fake fur.

7.2 Description of Environmental Impacts

(a) Ozone layer depletion:

Most atmospheric ozone is found at an altitude of around 15–30 km and this part of the atmosphere is therefore known as the ozone layer. This layer absorbs much of the damaging ultraviolet radiation emitted by the sun. The ozone layer is depleted by a variety of gases like chlorofluorocarbons (CFCs), which results in decline of layer thickness (Bijleveld et al. 2011).

(b) Ionising radiation:

Ionising radiation results from the decay of radioactive atoms like those of uranium-235, krypton-85 and iodine-129. There are two types of ionising radiation: particle-type radiation (alpha radiation, beta radiation, neutrons, protons) and high-energy electromagnetic radiation (X-rays, gamma radiation). Ionising radiation can damage DNA and cause a variety of cancers (Bijleveld et al. 2011).

(c) Respiratory inorganics:

Respiratory effects resulting from winter smog caused by emissions of dust, sulphur and nitrogen oxides to air. Damage is expressed in disability-adjusted life years (DALY)/kg emission (Earthshift 2016).

(d) Carcinogens:

Carcinogenic affects due to emissions of carcinogenic substances to air, water and soil. Damage is expressed in disability-adjusted life years (DALY)/kg emission (Earthshift).

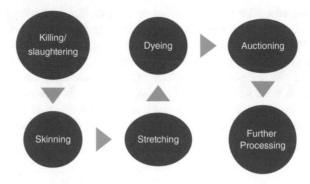

Fig. 1 Steps in natural fur production

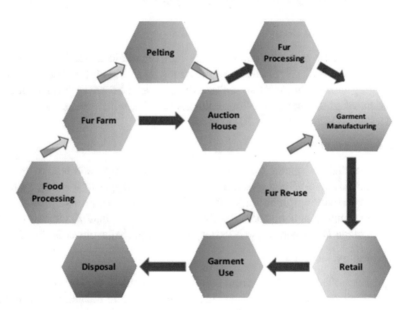

Fig. 2 Production, use and disposal stages of natural fur (DSS Management Consultants Inc. 2012)

Figures 1 and 2 show the flow of inputs and outputs associated with the life cycle of a natural fur and fake fur. For each stage, all of the major inputs and outputs are identified (Fig. 3).

The red arrows indicate the primary flow path among the process stages. The pink shaded arrows indicate operations that may occur as part of an integrated mink farm operation or that may take place off-site. The green shaded arrows indicate the potential for some of the product waste flow to be reused.

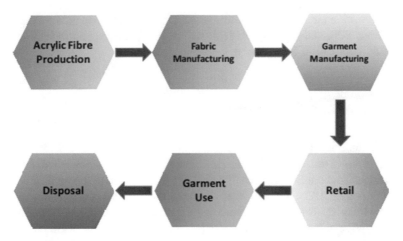

Fig. 3 Production, use and disposal stages of fake fur (DSS Management Consultants Inc. 2012)

8 Economic and Financial Considerations

The production of unprocessed fur tends to take place in developed countries, while the processing of fur and production of fur clothing take place in countries with a lower GDP per capita (Hansen 2014). The significant difference in wage costs, increasing globalisation and intense international competition mean that low technological and labour-intensive production moves to countries with low costs, something which is especially true for the fur sector. However, it should be mentioned that fur animal production has become very advanced, and in reality, the production of high-quality fur requires a range of skills.

In the 1990s Russia was the largest producer of fur in the world. But in Russia, the production costs to raise farmed fur animals have risen since the end of the Soviet Union, prior to which fur farmers had ready access to domestic fish and meat processing by-products. As a result of government reforms, the Russian Government no longer subsidises feed or offers easy credit terms. Feed costs, which represent the greatest cost of producing a pelt, rose as the infrastructure of industries which provided domestic fish and meat by-products were destroyed. Russian fur farmers thus incur greater production cost compared with other major farmed fur-producing countries (USITC 2004).

In Asia, Hong Kong and the mainland China are the largest producers and consumers of fur (HKTDC). According to Hong Kong Trade Development Council, under CEPA III (Closer Economic Partnership Arrangement), the mainland China agreed to give all products of Hong Kong origin, including fur items, tariff-free treatment from January 2006. The majority of Hong Kong's furriers have set up production facilities on the Chinese mainland amid higher production costs in Hong Kong. Still, many major sub-sectors of the fur industry, particularly sales and distribution, remain in Hong Kong.

9   Licenses and Regulations

Different countries have different regulations on fur farming. Fur farming and trapping follow the international agreements such as CITES (Convention on International Trade in Endangered Species), the Convention on Biological Diversity (CBD) and the IUCN (World Conservation Union). Provincial and territorial wildlife biologists establish regional biodiversity plans to ensure healthy wild furbearer populations (Fur Institute of Canada). Also, International Humane Trapping Standards (AIHTS) ensures that the animals trapped follow humane trapping standards.

In European Union laws exist on fur farming and animal welfare. 32.1 million animals are killed each year in EU for fur farming (ESDAW 2016). In EU the most commonly farmed animal species is the mink followed by the blue fox. Following are some of the rules outlined by the European Commission for fur farming:

(1) Housing conditions for animals caged must be of certain specifications. For example, 70 cm long by 40 cm wide and 45 cm high for minks.
(2) The quality and the composition of the feed must be controlled. Feed is mainly composed of fish and fish offal, poultry offal, slaughter house offal and cereal with mineral and vitamin ingredients.
(3) For fox and mink both open and closed buildings are used. Coypus are always housed outdoors and chinchillas indoors.
(4) Farmers and other persons responsible for the animals should be authorised to keep animals for fur production only if properly trained in all relevant aspects of their biology, welfare, and management.
(5) Daily inspection of animals must be maintained.
(6) Animals born in the wild should not be brought into farming conditions.
(7) Restraint devices should be used as little as possible.
(8) Killing of animals kept for fur production should be carried out only with humane methods. In particular chloral hydrate should not be used. Animals should be handled gently prior to killing.
(9) Mutilations of animals kept for fur production, e.g. detoothing, should be avoided

In USA and Canada, the Washington Convention (Convention on International Trade in Endangered Species of Wild Fauna and Flora (CITES)) restricts or prohibits the trade of certain species. Licenses must be obtained by fur farmers from their local and territorial government authorities for the protection of wildlife and environment.

10   Animal Welfare in Sustainability

What do we mean by animal welfare?

Animal welfare is the prevention of unnecessary animal suffering, i.e. ensuring a good quality of life and humane death (worldanimal.net). It comprises basically two elements: physical state and mental state.

(a) Physical state is the physical condition in which the animal lives and is coping with. For example, altered body functions due to lack of space and adaptability depending on the animal species can cause damage to their physical state (Dawkins 1980, 1990).

(b) Mental state is being explored more nowadays. Focussing on the stress levels and feelings that animals go through. As stated by Duncan (1991), the extent to which animals are aware of their internal state while performing behaviour known to be indicative of so-called states of suffering, such as fear, frustration and pain, which determine how much they are actually suffering.

10.1   Sustainable Animal-Based Products Versus Non-sustainable Animal-Based Products

It is important to note that the consumer consciousness has been increasing for the environmental conservation and animal rights. Consumers want to be sure of the ethicality within the fashion industry. For this, the fur industry started a voluntary labelling programme in 2007 called Origin Assured Label or OA™, which informs the consumers about the origins of products and regulation and standards governed during fur production (HKTDC). Furthermore, enhancement of the regulations on environmental conservation and animal protection is promoted. The biodegradable and less-pollutive fur during the production process is increasingly considered a sustainable material.

In the meantime, recycling fur has started to grab the attention of producers and consumers. HARRICANA PAR MARIOUCHE, a Canadian fur brand, for instance, has been saving more than 800,000 animals over the past 15 years by recycling old furs. In Canada, the Beautifully Canadian™ label is a guarantee that the garment is made from Canadian fur (Fur Institute of Canada 2016).

Another aspect of natural fur consumption is the use of by-products from the animal like in oils and fertilizers. For example, mink oil is used in medical and cosmetics industry. It is obtained by rendering of the mink fat which has been removed from the pelts destined for the fur industry (Wikipedia 2016).

10.2   WelFur

According to the Kopenhagen fur (2016), animal welfare assessment programmes like WelFur were initiated by the European fur sector in 2009. The pan-European implementation of WelFur began in 2015 with 10 European countries participating in the pilot scale. Overall, the WelFur system has three objectives:

1. To provide a reliable on-farm animal welfare assessment system based on scientifically proven measures and independent third-party assessments.
2. To improve animal welfare on European fur farms through analysing of the assessment data and education of the farmers.
3. To provide consumer transparency on the welfare status on European fur farms by publishing assessment data.

The welfare assessment protocols for fur farmed species (mink and fox) are developed by independent scientists at seven European universities (University of Eastern Finland (Department of Biosciences), MTT Agrifood Research, Finland (Animal Production Research), Aarhus University, Denmark (Department of Animal Health and Bioscience), Norwegian University of Life Sciences (Department of Animal and Agricultural Sciences), Swedish University of Agricultural Sciences (Department of Animal Environment and Health), University of Utrecht, the Netherlands (Department of Animals in Science and Society), French National Institute of Agronomic Research) and were published in 2013 and 2014. These protocols work as science-based 'manuals' for the third parties assessing the individual fur farm. Based on the principles of the European Commission funded Welfare Quality® project, the programme takes on a multi-facetted approach to animal welfare considering the parameters like positive and negative emotions, health, natural behaviour, housing system, feeding, human–animal relationship and the management of the farm.

11   Is There an Ethical Way of Consuming Animal-Based Luxury and Fashion Products?

In 1994, when PETA campaigned with five supermodels who sat naked on the floor and told the world "We'd rather go naked than wear fur", it reached millions of people and created successful awareness against the use of fur. However, some of the models from the campaign like Cindy Crawford, Naomi Campbell and Christy Turlington have been in recent years found to have posed and promoted fur garments.

According to Beard (2008), in 1980s consumerism became increasingly politicised in Britain, which was demonstrated through campaigns to prevent testing of cosmetic products on animals. While Body Shop was one of the firms most widely associated with the Against Animal Testing campaign, it was perhaps the appropriation of glamorous fashion photography in advertising by the pressure group Lynx that captured the public's attention in its quest to ban the use of fur.

As Andrew Bolton (2004) attests, "Acting as a form of "guilt politics," it urged women to reject fur in order to exhibit a morally as opposed to a materially superior status, thus giving birth to a new ideal of femininity, the moral or ethical woman".

Although Lynx failed to prevent the long-term continuation of using fur, as evidenced by its recent return to high fashion by designers such as Julien Macdonald, they succeeded in making the wearing of fur socially unacceptable to a wider audience, giving rise to the idea that being ethical could also be fashionable (Beard 2008). Eco-fashion therefore has emerged as another way for fashion brands to stand out in a highly overcrowded market.

In 2010, Ipsos Public Affairs, the global market research company, undertook a consumer survey about the image of fur farming in Germany, Belgium and the Netherlands. 1000 respondents from each country participated in the survey. According to this survey most participants had negative image about the fur industry as that of being cruel and that fake fur has less ecological impact (European Fur Information Centre 2016). This shows the declining demand of natural fur but not faux fur.

Within the sustainability paradigm pro-fur spokespersons have proclaimed there is not any difference between leather and fur as both the types of animal products utilise entire animal and not just the skin. However, the likeability of animal-based products in general is declining. The acceptance of leather by some and not fur is the key to understand ethical selective considerations made by consumers.

12 Discussion

12.1 The Debacle of Faux Fur as Sustainable

Sustainability is a long-term process considering the financial social, economic and other requirements of present and future generations. According to EPA (United States Environmental Protection Agency), sustainability is important in making sure that we have and will continue to have the water, materials and resources to protect human health and our environment. Many industry experts in the field of fashion and luxury proclaim their brands as sustainable due to the use of faux fur in their products. But is faux fur really sustainable?

Despite the fact that faux fur is saving animals from being slaughtered, it is not completely sustainable. The primary reason being that faux fur is derived from petroleum products which are non-renewable and non-biodegradable. Unlike the natural fur the faux fur does not last as long and is not durable. The other stance being that of fast fashion where faux fur fashion might come and go with new designs and patterns.

12.2   Where the Future Lies?

We outline some of the future forecast based on our research and current findings of the fur industry.

1. Brands will adopt fur and faux fur fashion to cater the different types of consumers. For example, luxury and fashion brand like Prada serves both the consumers against and pro fur by launching products with natural and faux fur in the market. Increasing demand of fur thus is prevalent and market driven.
2. Anti-fur campaigns from organisations and animal rights groups like PETA, anti-fur coalition, make fur history, born-free USA, etc., are becoming increasingly popular through the internet and the social media, which is helpful in spreading information about the sufferings and unethical practices of the fur industry. In addition, it makes consumers feel involved in contributing to sustainable practices.
3. As it is getting common on social media for celebrity fans to make a mountain of a mole, celebrities could enhance or damage their reputation based on their opinions on the use of fur.
4. Status will continue to drive the demand of fur, even in countries like China, India and Brazil where winter months are few.
5. Adapting to sustainable lifestyle will be more prevalent amongst the educated consumers, which does not necessarily mean abstinence from faux fur.
6. Consumer mainly wear fur in colder regions due to extreme cold temperatures. Therefore, it is an economic investment for certain consumer groups to buy natural fur.

Sales in the fur industry are driven by two kinds of consumers: the old consumers and the young ones under 40 yrs (Bukszpan 2015). The prime reason is being that of rewarding oneself with luxury and adhering to the latest fashion. Anti-fur campaigns like from PETA have hurt the sales of the fur industry during the 1990s and still continue. However, the haute couture fashion and celebrity endorsements have contributed to the recent increase. The challenges that lie ahead for the fur industry are

- Global warming and the ecological and environmental impact from the use of natural and faux fur.
- Increased competitiveness through production in developing countries like China.
- Increased regulations and strictness on the fur farming conditions in developed countries.

On the grounds of ethicality and morality, killing of animals for their skin be it leather or fur cannot be considered a sustainable practice. Nevertheless, one cannot be sustainable if the act involves killing a living being. Even though the regulations of killing fur animals humanely have been formed, it does not prove that killing in itself is humane.

A clear standpoint on weather natural fur or faux fur is superior is inevitable due to different subjective views on ethicality and sustainability. As famously quoted by Karl Lagerfeld—"In a meat-eating world, wearing leather for shoes and clothes and even handbags, the discussion of fur is childish". With many fashionistas and designers aspiring to be the Karl Lagerfeld of tomorrow, the use of fur in the luxury and fashion industry is to stay but differ from market to market.

References

Avnis, J. (2015). There's actually a way to feel good about wearing fur. http://qz.com/356854/theres-actually-a-way-to-feel-good-about-fur/ Accessed February 3, 2016.

BASF. (2010). *Pocket book for the leather technologist*, 4th Ed. Revised and enlarged. www.basf.com/leather. Accessed February 2, 2016.

Bearden, W. O., Netemeyer, R. G., & Teel, J. E. (1989). Measurement of consumer susceptibility to interpersonal influence. *Journal of Consumer Research*, pp. 473–481

Beard, N. D. (2008). The branding of ethical fashion and the consumer: A luxury niche or mass-market reality? *Fashion Theory, 12*(4), 447–467.

Bijleveld, M., Korteland, M., & Sevenster, M. (2011). The environmental impact of mink fur production Delft, CE Delft. Available via www.cedelft.eu. Accessed March 11, 2016.

Bolton, A. (2004). *Wild: Fashion untamed*. New York: Metropolitan Museum of Art.

Bukszpan, D. (2015). How fur became a fashion favorite again. http://fortune.com/2015/06/28/fur-fashion-comeback/. Accessed February 15, 2016.

Connecticut Furs Inc. (2010). Making a fur garment. http://www.ctfurs.com/garment.htm Accessed February 12, 2016.

Dawkins, M. S. (1980). *Animal suffering: The science of animal welfare*. London: Chapman and Hall.

Dawkins, M. S. (1990). From an animal's point of view: Motivation, fitness and animal welfare. *Behavioral and Brain Sciences, 13*, 1–61.

DSS Management Consultants Inc. (2012). Report submitted to IFTF. Available via http://www.wearefur.com/sites/all/themes/iftf/pdf/LCA_final_report.pdf. Accessed March 22, 2016

Duncan, I. J., & Petherick, J. C. (1991). The implications of cognitive processes for animal welfare. *Journal of Animal Science, 69*, 5017–5022.

Earthshift. (2016). http://www.earthshift.com/software/simapro/eco99. Accessed March 1, 2016.

ESDAW®—European Society of Dog and Animal Welfare. (2016). www.esdaw.eu Accessed March 5, 2016.

European Fur Information Center. (2016). www.furinformationcenter.eu. Accessed February 10, 2016.

Fur Institute of Canada. (2016). www.fur.ca. Accessed March 20, 2016.

Gagnon Thompson, S. C., & Barton, M. A. (1994). Ecocentric and anthropocentric attitudes toward the environment. *Journal of Environmental Psychology, 14*, 149–157.

Gintis, H., Bowles, S., Boyd, R., & Fehr, E. (2007). Explaining altruistic behavior in humans. In R. Dunbar & L. Barrett (Eds.), *Handbook of evolutionary psychology* (pp. 605–620). Oxford, England: Oxford University Press.

Griskevicius, V., Tybur, J. M., & Van den Bergh, B. (2010). Going green to be seen. *Journal of Personality and Social Psychology, 98*(3), 392–404. doi:10.1037/a0017346.

Gyetvai, Z. (2012). G-28 (buildings)/G-26 (products) choice of environmental indicators—Complete LCA. http://www.eebguide.eu/?p=3524. Accessed May 14, 2016.

Hansen, H. O. (2014). The global fur industry: Trends, globalization and specialization. *Journal of Agricultural Science and Technology. A4*, 543–551. ISSN 1939-1250.

Hansen, N. E., Creutzberg, A., & Simonsen, H. B. (1991). Killing of mink (Mustela vison) by means of carbon dioxide (CO_2), carbon mono-oxide (CO) and Nitrogen (N_2). *British Veterinary Journal, 147*, 140–146.

Hong Kong Trade Development Council (HKTDC). (2016). www.hktdc.com. Accessed March 5, 2016.

International Fur Federation (IFF). (2016). www.wearefur.com. Accessed March 9, 2016.

ISO 14040. (2006). *Environmental management. Life cycle assessment—Principles and framework*. Geneve: International Organisation for Standardisation (ISO).

ISO 14044. (2006). *Environmental management. Life cycle assessment—Requirements and guidelines*. Geneve: International Organisation for Standardisation (ISO).

Knight, D. K., & Kim, E. Y. (2007). Japanese consumers need for uniqueness: Effects on brand perceptions and purchase intention. *Journal of Fashion Marketing and Management., 11*(2), 270–280.

Kopenhagenfur. (2016). www.kopenhagenfur.com. Accessed January 30, 2016.

Miller, G. F. (2000). *The mating mind: How sexual choice shaped the evolution of human nature*. New York, NY: Doubleday.

Orange, R. (2014). Fur trade booms, fuelled by China—But bubble may be about to burst. Available via http://www.theguardian.com/fashion/2014/apr/19/fur-trade-mink-peta-china-bubble

Peterson, L. A. (2010). Detailed discussion of fur animals and fur production. Michigan State University College of Law. Available via https://www.animallaw.info/article/detailed-discussion-fur-animals-and-fur-production#s1. Accessed March 12, 2016.

Rokeach, M. (1973). *The nature of human values*. New York: Free Press.

Sojuzpushnina. (2016). http://www.sojuzpushnina.ru/en/s/55/. Accessed March 22, 2016.

Ruvio, A., Shoham, A., & Brencic, M. M. (2008). Consumers' need for uniqueness: Short-form scale development and cross cultural validation. *International Marketing Review, 25*(1), 33–53.

Ryan, H. Z. (2008). *Uniqueness and innovativeness: A look at controversial men's fashion products*. Perth: Curtin Business School, Curtin University of Technology.

Tian, K. T., Bearden, W. O., & Hunter, G. L. (2001). Consumers need for uniqueness: Scale development and validation. *Journal of Consumer Research, 28*, 50–66.

USEPA. United States Environmental Protection Agency. (2016). Defining life cycle assessment. http://www.gdrc.org/uem/lca/lca-define.html. Accessed May 12, 2016.

USITC. United states International Trade Commission. (2004). Industry and trade summary. Publication number 3666, Furskins.

Williams, A. S. (2009). Life cycle analysis: A step by step approach. ISTC Reports. TR-040. http://www.istc.illinois.edu/info/library_docs/tr/tr40.pdf. Accessed March 14, 2016.

Wikipedia. (2016). https://en.wikipedia.org/wiki/Mink_oil. Accessed March 12, 2016.

Zahavi, A. (1975). Mate selection: Selection for a handicap. *Journal of Theoretical Biology, 53*, 205–214.

Connectivity, Understanding and Empathy: How a Lack of Consumer Knowledge of the Fashion Supply Chain Is Influencing Socially Responsible Fashion Purchasing

Alana M. James and Bruce Montgomery

Abstract Consumer knowledge of the clothing supply chain remains minimal, with the majority of fashion customers having very little knowledge to the origin of their clothing purchases. Whilst they remain very familiar with the retail environment, the journey any one item of clothing goes through to reach the point of sale eludes them. Referred to as the consumer knowledge barrier, it is this lack of knowledge that is said to be influencing their socially responsible purchasing behaviour. The supply chain remains a complex process, however, with an increased lack of transparency, how consumers can obtain additional information about this remains a problem. Whilst consumers continue to be uninformed their power becomes meaningless, as they are unable to make informed purchasing decisions. Knowledge allows the consumer to choose where to shop, and where to avoid, in relation to their values. It is becoming more common to see retailers now engaging with corporate social responsibility as part of their everyday business practices. The level of engagement, however, remains varied with some companies being much more proactive in developing a strategy to help them move to more responsible practices. It is the communication of this strategy that allows retailers to engage consumers in these practices, informing them of such issues in the process. The adoption of this attitude promotes the linking of their consumers with the supply chain, taking a more transparent approach to business. The connection of the consumer with the supply chain not only increases their knowledge of ethical and sustainable issues in fashion but also aids in the creation of empathy and understanding with the social side of manufacturing. Currently, consumers are disconnected with behind the scenes of the fashion industry and cannot relate to the individual who produced the clothing they choose to buy. Through retailers creating

A.M. James (✉)
School of Textiles and Design, Heriot-Watt University, Galashiels, UK
e-mail: a.james@hw.ac.uk

B. Montgomery
Faculty of Arts, Design and Social Sciences, Northumbria University,
Newcastle upon Tyne, UK
e-mail: Bruce.montgomery@northumbria.ac.uk

© Springer Science+Business Media Singapore 2017
S.S. Muthu (ed.), *Textiles and Clothing Sustainability*,
Textile Science and Clothing Technology, DOI 10.1007/978-981-10-2131-2_3

this connection with the consumers and the supply chain they stop acting as the middleman barrier and begin to adopt a more holistic approach to their responsible business practices. This consequently will help in the consumer making more informed and responsible purchasing decisions, transferring some of the power to influence the direction of the industry's future back with those who buy into it.

Keywords Corporate social responsibility (CSR) · CSR communication · Ethical fashion purchasing

1 Background and Context

1.1 Introduction

Fashion is vast industry accounting for almost one-third of the world's employment, from the farmers growing the raw fibre right through to the retail sales assistants. In the UK alone, the industry is said to be worth over £21 billion with approximately six billion items owned, equating to an average of 100 items per person (Compare My Spend 2012). Sustaining this huge industry of employment is the fashion consumer, with the average customer spending £1232 on clothing alone each year (Kirk 2015). Despite this huge consumer buy into this obviously popular industry, the average consumer knows relatively very little about the origins of the garments they purchase and how they have been made. This lack of knowledge of the journey that garments take from fibre to customer is often reflected with a consumer lack of engagement with social responsibility in their supply chain. A lack of basic knowledge at this point does not allow a consumer to engage with ethical or sustainable issues as they first and foremost do not possess the foundation knowledge and context for which these issues are pertinent. The garment supply chain is a long and complex process, however, the generic consumer knows very little of even the most basic of processes undertaken in the production of fashion goods. This lack of knowledge is said to be due to a number of reasons will be discussed in more depth later in the chapter.

How and where consumers obtain information regarding the supply chain has been highlighted as problematic. Despite an increase in media coverage in recent years, if a consumer does not actively seek to obtain information to better inform them of the story behind a garment, there is very little provision of this information. Media coverage tends to occur when an event or breach of regulations has taken place. This reactive approach whilst in the very nature of the press, does not aid in the average consumer gaining knowledge to better inform their purchasing decisions they are making. The fact that information is not readily available to consumers means that unless consumers are very interested in obtaining this information for themselves, they will probably not go actively looking. Without this initial basic knowledge of the supply chain, consumers will not be able to fully appreciate the social responsibility required during the manufacture of garments.

The chapter will begin by outlining not only the importance of consumer knowledge of social responsibility in fashion but also the issues currently preventing this. Factors such as product segregation and a lack of available information are just a couple of examples that will be discussed in more detail. The supply chain will be discussed in Sect. 2.1, with the complexities of this process detailed and the benefits of further transparency in the pursuit of moving the industry to a more socially responsible future. Section 3.1 will begin to explore the consumer–retailer relationship in more detail and how companies can engage themselves in social responsibility through the integration of a corporate social responsibility (CSR) strategy into their business practices. Furthermore the communication of this strategy can also engage the consumer in these responsible practices, informing and increasing knowledge of issues surround ethics and sustainability. A CSR communication study will be presented where the authors took a case study approach working with five high street retailers. This study investigated the existing communication utilised by the retailers in informing their customers of their responsible practices. The type of information available and the methods of communication remained the focus of the study.

The chapter will conclude with the case for increased transparency in the supply chain being discussed. The authors will argue this can be achieved through increased CSR communication. The retailer remains in a powerful position having influence on both the purchasing process in their relationship with the consumer but also the supply chain with their manufacturing suppliers. It is this influence that they must use in order for this transparency to increase and for the consumer to feel more informed and connected to where their clothes come from. This connectivity will aid in the creation of knowledge and empathy with the consumer and consequently influence their socially responsible purchasing behaviour.

1.1.1 The Consumer Knowledge Barrier

This lack of knowledge and awareness on the part of the consumer is posing a problem in aiding the industry move forward to a more socially responsible future. However, there are several factors, which are currently preventing consumers from engaging more with the supply chain and furthermore the understanding of social issues, which occur during this time. The factors preventing further consumer knowledge and acting as knowledge barriers can be summarised as follows:

- A lack of knowledge provision for the average consumer
- Interested consumers not knowing where to go to obtain information
- Information provided not being engaging or interesting to the consumer
- Retailers not readily making the information available.

The saturated fashion market offers more choice to the consumer than ever before, facilitating the wants and needs of consumers regardless of their preference. The ethical market however is the one sector of the industry that remains uncatered for, with accessibility to socially responsible fashion still being the exception rather than the rule. The ethical market accounts for just over 1 % of the fashion market (Niinimaki 2010), which could be questioned if this small market share facilitates the needs of the socially responsible consumers. This juxtaposition between an oversaturated fashion market offering consumers wide array of choices and the lack of provision of socially responsible apparel remains an issue preventing change for a more sustainable future.

It could be questioned, however, if this segregation is the right way forward for the industry in the context of ethics and sustainability. This approach continues to give the impression that a more responsible fashion supply chain is an option which consumers can choose to buy into, in preference to the way business is always conducted. This separated market is also emphasised further by some retailers, where certain ranges are highlighted to be adopting more ethical and sustainable approaches to garment manufacture. An example of this would be the Eco-Conscious collection at H&M which provides customers with a nature-inspired collection utilising high percentages of recycled polyester fibres in their fabrics. Whilst this is a commendable contribution to the impact of creating virgin fibres, this still segregates the collection from the mainstream collections. Savvy consumers may also begin to question why certain collections are being separated for being more responsible than others, with questions being raised regarding the credentials of the more mainstream collections being offered. Another example of this approach would be retailers and brands that use organic Fairtrade cotton in their basic apparel ranges. Often highlighted on the garment swing tags, retailers have begun using more responsible sources of cotton such as organic (the elimination of pesticides during the growth of the raw fibre) and Fairtrade (where a premium is paid to enable giving something back to the local cotton sourcing community). The use of these alternative types of cotton is often only used in basic ranges, simple garment designs often using basic forms of cotton textiles. Again this is a commendable approach to the development of some garment ranges, but this needs to be included in a wider range of products available. More complex garments, involving more fashion garment detailing and a wider range of fabrics need to begin to be included in this approach, bringing organic and Fairtrade cotton more into the mainstream fashion ranges for brand and retailers.

Whilst retailers continue this separation of socially and environmentally responsible collections, this two-market approach will exist. In preference to highlighting socially compliant collections, retailers and brands should begin to look to ways they can incorporate more ethical and sustainable practices into their everyday way of doing business. Examples of this could be the increased use of organic cotton in their garments, working more closely with manufacturers in their supply chain and the development of reuse or recycle schemes at the end of the product life cycle. The incorporation of social and environmental responsibility into the core business for fashion retailers would eliminate the element of choice for consumers, who could place their trust in certain retailers opposed to being limited to certain collections.

This element of control on the part of the retailer is again a heavily debated topic in the industry. Questions have been raised into who controls the industry; the consumers demanding certain goods to be available or does the retailer ultimately control what the consumers can purchase. This two-sided debate has continued for many years with the consumer's purchasing behaviour said to be a powerful tool in the success or decline of specific retailers or brands. It is believed that the very act of purchasing is a consumer's expression of their own individual judgment and values, however this can work both ways, with consumers also choosing not to shop somewhere due to disagreeing with a companies behaviour (Smith 1995).

This chapter will focus on the issues currently preventing further consumer knowledge of the fashion supply chain and the social and environmental issues involved. The vital relationship between the consumer and the retailer will be explored and how improved communication between the two can aid in a more informed consumer. The engagement in CSR from high street fashion retailers will be discussed, with the methods of communication investigated through a research case study. To conclude the chapter, the key problems facing an informed consumer will be presented followed by a series of recommendations to help the connectivity of the consumer with the garment supply chain. This development is believed to be the key in creating understanding and empathy in order to influence socially responsible fashion purchasing.

2   The Fashion Supply Chain

2.1   The Basic Supply Chain

The supply chain is a long and complex process, with nearly every fashion product being produced through a different variation of this process. During the production of even the most simple of garments, the process undertaken from start of manufacture to consumer can be a long and varied journey. When considering the supply chain at a basic level, these differences predominantly occur due to variables individual to the product itself. This can range from fabric type, method of fabric construction, dyeing methods, embellishment techniques, garment construction and the finishing processes utilised. Taking the supply chain in its simplest form starts with the acquisition of the fibre, if natural then this will be farmed, if man-made then this will be produced. This fibre then needs to be spun into a yarn during the spinning process, followed closely by the knitting or weaving process which will take the yarn and create a piece of flat fabric which can then be used to make the garment. The dying of the final piece of fabric can occur either in yarn or fabric format dependent of the design of the textile. Any printing or embellishment may also take place at this stage; however, prior to this being utilised in the garment manufacture process the fabric must be finished to ensure durability for the user. This fabric is then cut and manufactured into a final product before being retailed and sold to the consumer (Fig. 1).

When applying this simplified supply chain to example garments, there are additional variables (fabric type, method of fabric construction, dyeing methods, embellishment techniques, garment construction and finishing processes) that can be identified. An example of variation in the supply chain is that of an embellished, woven, polyester blouse. The man-made nature of the composite fabric means that the fibre is not grown or picked but rather extruded in its yarn format. This is due to polyester being a plastic-based fabric taking its origin from oil. The construction of the fabric also adapts the supply chain process, as the fabric is woven, looms are required opposed to knitting machines. The embellishment on the blouse also adds a further stage to the already long and complex process. Dependent on the retailer and nature of the embellishment, this can either be applied by hand or machinery. Monsoon, for example acknowledges the complex nature of a lot of its embroidery work and employs homeworkers in India to conduct a lot of the intricate work (Ethical Trading Initiative 2010). In comparison to machine embroidery, for example this obviously takes much longer to execute and would therefore be unsuitable for some markets such as fast fashion.

The differences created in the supply chain are often due to the utilisation of a number of different suppliers in varying regions of the world. One garment can

Fig. 1 The basic garment supply chain. *Source* Authors

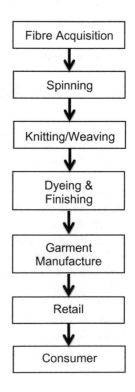

travel thousands of miles during production before reaching the customer, with the different stages of the supply chain often being in different geographical locations. For example, a basic Jersey cotton T-shirt, the raw cotton fibre may have been sourced from an agricultural region of India, the processing of this raw fibre (cleaning, bleaching) and the spinning of the raw fibre to turn this into a useable yarn may occur in Sri Lanka. Due to the nature of cotton, this yarn will be in a natural colour or greige (industry terminology), indicating that it has not been dyed during the production process to this point. This process may also occur in Sri Lanka, where the resulting yarn will be the colour of the desired product. With the yarn then making its next stage of the journey where it is shipped to Bangladesh where it is knitted into a piece of fabric. This fabric may then be shipped to Morocco where the CMT (cut make and trim) of the final garment will take place. This final garment may then be parcelled and packaged up and sent to its final destination of the US to be sold to the mid-level womenswear market. Although it is to be acknowledged that this is only an example supply chain, it is to demonstrate that from fibre to customer can span thousands of miles during this long and complex journey.

2.1.1 Supply Chain Transparency

As has been demonstrated the supply chain is not a simple journey with many complex stages having to be undertaken in the manufacture of a garment. These supply chains are dictated by the brand or retailer and again varies from company to company. However, what retailers also control is how much information they reveal to the public about their supply chains and the methods utilised in the production of their products. This is often referred to as transparency and refers to the amount of information revealed about the suppliers and methods utilised to produce garments. Again this varies greatly from company to company, with some stating sometimes no information at all or simply just the country of manufacture, ranging to some companies who publish factory lists including names and addresses of the supplier factories used. This level of transparency, however, could leave the company open to criticism due to the open and honest nature of the information being provided. The majority of companies reveal some information but refrain from disclosing the details of their supply chain to avoid being in a media expose should something go wrong. Again due to the long and complex nature of the supply chain, the control that a company has on ensuring certain standards are met throughout can be difficult. Most retailers, especially in the production of fashion goods produce a code of conduct, which is a set of rules and standards that must be met by suppliers. The extent of these rules does vary from company to company and are a reflection of their values or CSR philosophy.

The benefits seen from transparent business are multiple, including the creation of brand trust in consumers. Disclosing details of manufacturing supply chains is an outward indication that as a company they have nothing to hide or anything that they do not want the public to know. This level of transparency can aid in the creation of trust as consumers will acknowledge and respect this decision even in they do not fully read or understand the details provided. This building of trust could eliminate in turn the element of choice provided to the consumer between companies who reveal their supply chain details and those who do not. This obviously relies on consumers possessing an existing level of knowledge regarding the production of garments they purchase but also knowledge of the issue which can occur during this long and complex process. This approach to transparency in garment manufacture can also help the company to build a reputation of being a responsible company. Again the message being portrayed to the public is that they have nothing to hide and that they are confident that no compliance issues are occurring in their supply chain.

When referring to consumer awareness of where and how fashion products are made, transparent business practice is a tool, which can successfully aid in the growth of this knowledge. Through making the details of supply chains available in the public arena, consumers will have access to relevant information, which may engage them in knowing more about their product choices. This approach will also allow the more interested consumers to again make choices in light of what they know.

An example of a company successfully carrying out transparent business activities would be Patagonia. Producing a range of outdoor clothing they disclose details of their manufacturing supply chain both on their website and through media outreach. Patagonia is also unique in the sense that they actively discourage their customers from purchasing unnecessary products from them. In their 2011 pre-black Friday advertisement campaign they revealed a simple slogan; 'don't buy what you don't need'. This was reinforced with another advertisement stating 'don't buy this jacket' (Patagonia 2011). This was an acknowledgement from the company to the current pace of the industry and to actively discourage overconsumption. This brave move again boosted the overall reputation of the company whilst reflecting their business values and philosophies.

Initiatives to boost transparency and to help retailers engage in this approach to business have in the past been developed. One of the most widely used of these is Made-by, a non-profit organisation who aid companies in a track and trace scheme of their supply chain. Indicated through the inclusion of a small blue button, engagement in the scheme allows consumers to log into an online system to see where and how their chosen garments have been produced. This initiative has been utilised by large fashion retailers such as Ted Baker, G-Star, Vivobarefoot and Haikure (Made-By 2016). Despite this approach not yet becoming mainstream, it is companies such as those discussed who are leading the field in its move towards a greater transparency in the fashion supply chain.

As mentioned previously, however, controlling supply chains in their long and complex nature is not easy, with companies often coming up against compliance issues as a result. There are several different systems and approaches undertaken by companies in order to aid in avoiding these potential issues. These can help control and monitor supply chains. This is the first stage of controlling their supply chains, implementing a code of conduct where rules and standards have to be adhered to by a supplier if they wish to continue working with the company. Many of these rules are social orientated meaning that they refer to the people working within the supply chain and the working environments in which the company's products are produced. Social compliance in the supply chain is paramount due to the significant consequences of non-compliance. Breeches of these rules have been witnessed in recent years with several social disasters occurring such as the collapse of the Rana Plaza complex in 2013 and the fire at the Spectrum Sweater factory in 2008. Both of these fatal incidents occurred due to the non-compliance of rules and regulations in the garment supply chains.

It is the implementation of these rules, however, when it can become difficult. The most popular tool implemented by most companies is auditing. Defined as a checklist for safety rules and regulations, this is where relevant company employees conduct factory visits to suppliers to ensure that their factory code of conduct is being adhered to. Issues observed include the health and safety of the working environment, working hours, provision of paid overtime and employment records. Again these rules and regulations vary from company to company and can often mean one factory has to satisfy several company audits due to their engagement with many. In addition to planned audits, companies may also conduct on-the-spot checks to make sure that advance preparation is not hiding any potential issues. Many companies also implement an online electronic system, which allows a two-way communication process between the company and the supplier. This often automated system allows the company to follow products through their supply chain and monitor where non-compliance issues may occur. A change in company attitudes in recent years has seen a more proactive approach to solving such issues. During interviews conducted by the author in 2010, two major high street retailers discussed how they aim to work with a supplier should issues occur in following their code of conduct. Whilst non-compliance is treated very seriously, companies are beginning to build stronger relationships with suppliers to implement strategies to prevent further issues happening again in the future. This approach is often taken in preference to a severance of relationships where new relationships would need to be developed covering old ground previously undertaken with existing suppliers. This closer working relationship is now seen as a preferable response to non-compliance issues to only help build the strength in working relationships in the supply chain between two key stakeholders; retailers and their suppliers.

The hiding of information is unfortunately a common occurrence during the auditing process with some factories passing some of their large and pressured workloads onto third-party factories. The key issue with this approach is the lack of control and regulation companies have over the working standards of these additional factories. Often labelled as 'shadow factories' (Harney 2008) the unregulation of such factories can have major consequences in compliance of their code of conduct. These issues can include breaches of health and safety standards such as, lack of fire exits and excessive working hours. All of which cannot be controlled with third-party suppliers or shadow factories often resulting in non-responsibility in the supply chain. When the company code of conduct is not complied with is when media expose can happen. The risk of this ironically is one reason which discourages more companies from utilising transparent business practices.

3 The Consumer–Retailer Relationship

3.1 The Power of the Consumer

The relationship between the consumer and the retailer is often approached with some degree of cynicism. Consumers tend to have a sense of distrust towards the retailer often being centred on money and the price that is being asked for a certain product or goods. This feeling is cemented within the purchasing process, where the consumer will initially search for a product that fulfils their need or want identified. Following this the consumer will begin to scope out the market, searching often for the best product at the best price available. The difference in price between retailers helps to promote this distrust, with some retailers offering the same products for much higher prices. From a consumer perspective this appears to be the retailer simply wanting to make more profit from sales, however, what the consumer often does not understand is the business-orientated rationale behind the price provided to the consumer. The business model for example can vary the price of the same product significantly. If one retailer is able to purchase a much higher amount of that original product from suppliers, then a cheaper price may be obtained. However, on many occasions the consumer instinct may be correct, with some retailers having a higher markup on their products in order to either make more profit or cover high overheads accumulated through their style of retail.

This balance of power during the purchasing process has been identified by companies who are actively working on creating consumer trust. This creation of trust, however, is not easily developed and takes a lot of work and time on the part of the retailer. Despite the time and effort taken in order to build this trust, it is very easily lost again due to many different reasons. If the consumer feels that they are not being offered a fair price, quality goods/services or treated with mutual respect during the purchasing process, this relationship can easily be tarnished with trust diminishing as a result. The coverage a company receives in the media can also have a lot of effect on the perception a consumer may have of a brand or retailer. For example if a company is sold to a large multinational or global conglomerate, the trust established may also be damaged due to the unknown values of this new owner. With regards to social responsibility the media is again a powerful tool used to inform consumers about the retailers with which they choose to shop. Many fashion companies have in the past been involved in media exposes regarding unsavoury labour conditions including workshop conditions and child labour. The biggest expose seen in the fashion industry dates back to the 1990s where Nike was exposed for producing their garments using child labour practices. This was heavily covered by the media who informed the public that Nike products were being produced by 13 and 14-year-old children, working 20 h a day in garment factories in Honduras (Strom 1996). In 2007 fashion brand Gap was also exposed for its use of child labour in India, where children as young as 10 were used to embroider logos and decorative pieces onto Gap clothing (Gentleman 2007). However, since these events both retailers have been working hard on reshaping their values towards social responsibility in a bid to aid the growth of brand trust with their customers.

Fast fashion retailer Primark was found to be one of the companies producing garments in the Rana Plaza garment factories, which collapsed in 2013 due to breaches of building safety regulations (Hartley-Brewer 2015). With consumer response being similar to that seen in 2007 sales were thought to have declined, however, this was quite the contrary with annual profits in November 2013 seeing an increase of 44 % to £514 billion, equating to £11.7 million per trading day (Hawkes 2013). Despite the general public relying on press for the latest information, the authenticity and accuracy of stories must also be questioned, often being further elaborated or dramatised for the increased effect of text-based news.

Although the consumer cannot control or regulate the actions or business choices of companies, they do have the power to choose where they shop as a reflection of the values of the retailer. This choice of where they purchase is effectively a vote, indicating that the consumer is happy to spend their money with that particular company. This is similar to that seen during a boycott, where consumers actively choose to not purchase with a particular company due to disagreeing with their business practices. This element of choice, however, does rely on the consumer being knowledgeable enough to make an informed purchasing decision. The accuracy of the information and the reliance of the source providing the information can also effect the decisions made by consumers. In addition to consumer choice, pressures on buying intentions created in peer and social circles can also be a powerful influential tool in the choice of purchasing retailers.

The power of the consumer has been identified in the market of food purchasing, where consumers have demanded that alternatives are made available. The response to this is evident with large quantities of food ranges now providing organic alternatives and goods such as coffee, chocolate and bananas offering Fairtrade options to customers. These alternatives and choices have become widely accepted in the food industry, however, this has not seen to be spreading to the fashion industry. Even the cosmetics market now provides more socially responsible alternatives to the mass market, with shops such as Body Shop and Lush including ethical and sustainable credentials in their USP (unique selling point). With industries such as food and cosmetics moving forward, the fashion industry does not appear to be keeping pace with consumer demand, which is seen as a direct influence of increased knowledge and awareness of ethical and sustainable issues. The increased media coverage of relevant issues is said to be responsible for the growth of consumer interest and should in turn see retailers responding in the way seen in other markets. In support of this argument it can be questioned if it is retail giants who dictate how consumers purchase goods (Fox 2005) or if consumers really can begin to demand the type of goods available or ultimately the methods used during the manufacturing supply chain.

3.1.1 Corporate Social Responsibility (CSR)

In addition to consumers engaging with ethics and sustainability in fashion, retailers and brands alike also adopt certain values into their business philosophies. As previously discussed, the fashion supply chain is a long and complex process, which can be prone to issues of social responsibility in one form or another. However, as consumers become more interested and engaged in such issues surrounding the fashion products they choose to purchase, retailers need to also see the benefit of engagement (Wales et al. 2010). This engagement has been called CSR and can be defined as the process of assessing an organisation's impact on the environment and society. It is a commitment to improve community well-being through discretionary business practices (Kotler and Lee 2005). However, just as with the terms ethics and sustainability there is again no industry definition or standards and regulations for a company to work to (Ihlen et al. 2011). This again leads to complications with engagement as the term CSR can be used despite the level of action or engagement. Naturally, this lack of regulation allows companies to be engaged in CSR even if their commitment is minimal and will use the same terminology as companies who are very active.

The scope of activities covered by the term CSR is vast with as previously discussed the terminology covering all actions that are for the good of society or the environment. Some examples of company actions which contribute to CSR business practices include; the planting of trees in order to offset carbon emissions produced during the manufacturing of fashion products, the education of garment workers in basic literacy, numeracy and healthcare skills, the reduction of water use in the production of denim products and the recycling of garments coming to the end of their product life cycle. The list of methods of engagement is endless and again varies from company to company. This has lead to the belief that CSR should not adopt a one-fits-all type of approach, with engagement working for some companies, under some circumstances, some of the time (Vogel 2005).

Due to the complexities with terminology and definition, some brands and retailers have chosen to use alternative names for their actions such as commitments, values and philosophies. This development often comes hand in hand with companies engaging well in the CSR process, indicating their level of engagement and commitment. This can also show that the companies have begun to embed these values into their core business practices instead of just additional ranges amongst their usual offering to consumers. Adopting this approach means that companies are permanently adapting processes and practices to that of a more responsible business model. Some companies have gone as far as to brand their CSR commitments, often for ease of distinction and communication purposes. Examples of this would be Plan A at Marks and Spencer and L.O.V.E (Love our Values and Ethics) at Monsoon Accessorize.

Rationale for engaging with CSR is varied. This can include consumer expectation with heightened media coverage, knowledge and awareness. This consumer assumption again puts a certain amount of pressure on companies to be engaging in more responsible business practices. The increase of knowledge and awareness in recent times in consumers has had a consequential effect on the expectations of CSR engagement, with companies having to step up to this expectation of their customers. This relates directly to the reputation of the company, which again the inclusion of CSR is said to have a positive effect on retail brand image. It is also thought that there is a direct relationship between CSR engagement and financial performance (Zadek and Chapman 1998). A change in attitude towards CSR has also been witnessed in recent years, with perspectives adapting from that of profit sacrifice to profit gain (Burchell 2008). Also described as long-term profit maximising, companies are beginning to see the benefits of engagement through their finances as well as customer loyalty. Other benefits of CSR commitments include enhanced opportunities with innovation being said to being driven forward as a result (Wales et al. 2010). The level in which a company engages in a responsible business model begins to set them apart from competitors, indicating to customers the difference between themselves and their key competitors. This competitive advantage is again seen as a positive consequence of CSR engagement, which continues to sell the business case for engagement to businesses in all markets.

However, alongside positive consequences of responsible business practices comes scepticism and contradictory perspectives. The transparency of CSR practices have raised concern, with there often being a disparity between what is said to be being done and what is actually being done. This again relates back to brand trust, and the negative implications CSR engagement can have if this is found to be the case (Waddock and Googins 2011). This also promotes consumer distrust in the market with consumers often found to shut off when mistrust is found. This gap between what a company says they are doing in terms of CSR actions and what they are actually doing has been called *corporate hypocrisy*, with perceived hypocrisy said to be having a negative effect on consumer attitudes towards retailers (Wagner et al. 2009). This gap between CSR intentions and the translation of this into behaviour can be compared to the intention–behaviour gap found in the consumer purchasing process. This is where there is a stark difference between the purchasing intentions of a consumer and these intentions translating into consequential behaviour. This phenomena is often identified in social responsibility, with consumers often intending to purchase according to their morals, however, due to a number of reasons their final purchasing behaviour may not reflect their attitudes to social responsibility in fashion. This difference between what is said to be being done and what is actually being done is also seen with companies engaging in CSR (Aras and Crowther 2009; Cerin 2002; Wagner et al. 2009).

Due to consumers' purchasing behaviour often not seen as a reflection of their intentions or morals, companies often do not see a business case for engagement as profits do not reflect their level of commitment (Devinney 2009). If consumers were to shop by their beliefs and morals, this would not be the case. The industry as a whole also does not widely recognise or reward companies who are outperforming competitors in responsible business practices. This signals to companies there to be little point in engagement and adaptation to a more sustainable business model. Again terminology and standards becomes an issue and only encourages scepticism in CSR. The broad nature of the term results in it being used as an umbrella term, not indicating the type, depth or level of engagement from a company. As a consequence of this, some less reputable companies have began to pick and choose the types of CSR activities they become involved with, often choosing the cheaper and easier to implement strategies which is said to be of great concern to the industry (Kotler and Lee 2005).

3.1.2 CSR Communication

In addition to engaging with CSR activities, some companies like to also communicate what they are doing to a wider audience. Companies choose to communicate about what they are doing for a number of reasons: to inform key stakeholders, to interest and engage customers, to provide a public statement for NGOs (non-governmental organisations) and the press and to remain competitive in the market in comparison to other companies. However, the very act of communication of CSR values is thought to be making more information available to not only the public but also to decision-makers, resulting in further transparency in the supply chain. As previously mentioned the need for transparency in terms of social responsibility has never been needed more.

Just as with the level of engagement in CSR, different companies choose the different levels of publicising their values and actions with regards to environmental and social responsibility. The most basic of public engagement is annual reporting, which sees the companies issuing a report, which details their values and progress in terms of actions for the past year. Again at the most basic level this is seen as a business document to inform key stakeholders in the company, however, when utilised more effectively annual reporting can be a very effectively communication tool. Annual reporting is said to increase public trust in a company especially as the consumer can see what progress is being actively made by the company with regards to environmental and social responsibility. Public communication is also said to promote corporate accountability and avoid assumed PR (public relations) stunts through external verified reporting (Burchell 2008). False reporting of CSR developments have been termed *green washing* and can be defined as using CSR communications for the financial or reputational benefit to the company (Klein 2014). Whilst generally CSR engagement is perceived as a positive step for the company, the method of communication is crucial in its success. Critics have expressed concern over companies concentrating on strong methods of dissemination rather than the actual engagement itself. The varying methods of communication used by companies have also come under criticism, with the wide variety said to be difficult to compare and contrast across companies for parity. It is also feared that if companies are actively reporting, that they feel they are doing some good, even if that good is unsubstantiated and simply clever marketing (Burchall 2010). It is the pursued actions of the retailer, which will help them build trust through their outward, public communications. As a consequence of increased CSR communication, the trust created between consumer and retailers is changing. This is moving from the consumers trusting the retailer to that of wanting to know more and being shown the work that is going into implementing CSR strategies (Holme and Watts 1999).

CSR communication is vital in a consumer's perception of a brand or retailer, and can impact not only attitudes but also purchasing behaviour. Traditionally, this communication between a consumer and a retailer comes in the form of marketing principles, where retailers can deliver (often subliminal) messages regarding in-store purchasing. A variation of marketing principles is social marketing where behaviour is influenced for the good of society (Kotler and Lee 2005). However, it could be argued that marketing in its very nature is to encourage levels of consumption and materialism and thus going against the philosophy of ethics and sustainability in clothing (Miller 2009). It is thought, however, that the application of these principles could prompt an adaptation in fashion purchasing behaviour, significantly reducing levels of clothing consumption and consequently decreasing environmental impact and the social pressures in the fashion supply chain.

Social marketing can be defined as a process that applies marketing principles and techniques to create, communicate and deliver value in order to influence a target audience behaviours that benefit society as well as the target audience (Kotler and Lee 2005). Literature agrees with this, however, adds that benefits should be without financial profit to the marketer (Smith and Ward 2007). It is also thought that marketing should influence the voluntary behaviour (Andreasen 1995), whilst it is also thought that these principles should be applied to achieve specific behavioural goal awareness, knowledge and purchasing behaviour of fashion consumers. Retailers are already applying various communication methods in order to inform their customers of their social or environmental activities and intentions. Whilst marketing principles are applied in-store everyday, this communication tool could be utilised to influence the consumer behaviour for the good of society, whilst also practicing good business and remaining profitable.

As previously mentioned the methods to communicate CSR values and actions of a company is crucial and can be the difference between these values being positively or negatively perceived by the public. However, the consequences of increased communication are not always a positive one, with the wrong kind of communication methods often causing further scepticism (Lindgreen and Swaen 2010). Despite this, retailers are aiming to carefully select communication methods in order to demonstrate their involvement in a variety of CSR activities (Waddock and Googins 2011). The methods chosen are wide and varied and are often adapted to suit the business setup of each company. In addition to annual reporting, companies are now choosing to use innovative methods in order to get their message across with campaigns and initiatives often being their method of choice. These additional methods include;

- **CSR websites**—These are often additional micro-sites provided by the company taking the user away from the product purchasing site and moving to a more corporate site. The tools often include imagery, text, video and animation. This is also when branding of CSR is heavily practiced if the company takes this approach.
- **Social labelling**—This tool is predominantly used in-store where a customer can physically see and read a tag. This is an opportunity for retailers to inform their customers of any socially responsible factors applicable to the garment such as Fairtrade cotton or reminding them to wash at lower temperatures. The formats of social labelling seen in-store include swing tags and garment labels.
- **In-store marketing**—As seen with traditional in-store marketing, this can also be used to inform customers of the companies' CSR values and actions. This could take the form of point of sale marketing or large in-store visuals such as banners. Companies who choose to brand their CSR actions often use this tool to best communicate to consumers in the in-store environment.
- **Charity collaboration**—Many companies have chosen to collaborate with a charity for one of their commitments to CSR. Again this is publicised through many channels including in-store, online and direct public communication such as emails and text. This collaboration is often on the back of an initiative of campaign being run by the company.
- **Campaigns and Initiatives**—These are developed in order to encourage customer engagement and raise awareness of socially responsible actions being carried out. This is publicised through many channels including customer email, in-store advertising and website banners. As previously mentioned, these campaigns often have an interactive element, which again can help with the customer promotion and engagement.
- **Thank you campaigns**—This is a tool used after the launch of a campaign or initiative and is often used as not only a reward for the customer for their engagement but to also promote further engagement in the future. Through this direct correspondence, usually by email consumers are not only thanked for their contribution but also informed further of the CSR commitments and actions they have engaged with. Companies at this point also want to encourage positive repeat behaviour.
- **Database emails**—Purchasing and often non-purchasing customers are encouraged to provide an email address on many occasions during their interaction with the company. This is often stored on an internal database to utilise these contacts to inform customers of offers, promotions, events, etc. Again this is a tool used by companies to inform consumers and promote further engagement in their CSR values. This approach is direct which is positive but can be often overlooked due to the advertising nature of the email content.

- **Additional materials**—In addition to the methods used as previously discussed, companies also often take advantage of additional materials such a magazines, flyers and carrier bags. All of which can be utilised to inform and engage the consumer.

Regardless of the method of communication, a powerful tool in many of the methods discussed previously is the use of narrative. Many companies use story-telling as their main way to engage and keep the interest of their customers. The use of narrative is so powerful due to the familiarity of the structure of a story; people feel intrigued and want to find out more. When used effectively, companies can inform and engage consumers on an appropriate level regarding their CSR values and actions. The use of narrative is often accompanied with short video footage or eye-catching animations or graphics, which again reinforces the message conveyed in the story, engaging consumers to learn more. Examples of this include some retailers introducing their key suppliers or farmers to a wider audience by narrative or short stories aimed to get information across quickly. High street companies such as Marks and Spencer, Body Shop and Monsoon Accessorize all use this as an approach on their CSR websites. This aids consumers to feel more connected with the manufacturing supply chain using bite sized, easily accessible information. This approach is also adopted by The Fairtrade Foundation to give examples of their cotton farmers and workers; this personal approach again gives the consumer an insight into the work carried out behind the scenes in fashion.

Case Study—CSR Communication

The authors studied CSR communication during a short research project. This included the analysis of five high street fashion retailer websites for evidence of their approach to CSR communication. This study focused on two key areas;

- The type of information being communicated
- The method utilised to communicate CSR information

In addition to online websites, the in-store consumer experience was also explored for any CSR messages being communicated using visual in-store methods. The project included five high street retailers (company A, B, C D and E), all of which offered a womenswear provision to the UK high street market.

The results of the two communication studies (online and in-store) were dis-seminated into a matrix table, where each retailer could be analysed using a number of different criteria. For example on the online study, Company A's website would be examined for evidence of; *supplier ethical standards, factory lists, retailer code of conduct, audit process details, minimum wage, freedom of association* and *supply chain transparency.* Companies B, C, D and E would then go through the same process.

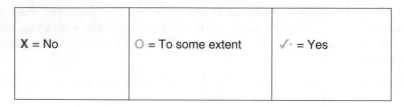

X = No	O = To some extent	✓ = Yes

Fig. 2 Traffic light matrix indicators

During the in-store data collection, the researcher visited each company store in turn, looking for evidence of CSR communication methods. Each of the following methods were analysed, with data recorded accordingly; *swing tags, care labels, carrier bags, leaflets/magazines, window displays, poster/advertisements* and *initiatives*.

The matrix-recording template allowed for the data collected to be categorised in accordance to the individual retailers level of compliance. A three-stage scale was used: *Yes, No* and *To some extent* and were visually recorded using colour coded icons to allow the reader to have a quick and easy overview of all the retailer case studies (Fig. 2).

Further details were also recorded, i.e. the type of CSR information featured on the carrier bags of company X. The matrix also accommodated a short summary of the store to be detailed again taking into consideration the original criteria identified.

Results—Online Analysis

This process was carried out using the retailer websites and where relevant, corresponding CSR micro-sites. The data collection process was done from the point of view of the consumer, accessing only information that was attainable by the general public.

The criteria developed for the online section, alongside working definitions of the terms were as follows:

- **Supplier Ethical Standards**—A set of minimum standards of compliance given to suppliers prior to manufacture
- **Factory Lists**—Specific information given regarding the factories used to manufacture products, i.e. names, location, etc.
- **Retailer Code of Conduct**—A set of values that the retailer works to reflecting the philosophy of the company as a brand
- **Audit Process Details**—The process undertaken by internal/external auditors to assess the ethical standards of a factory
- **Minimum Wage**—The minimum amount of money workers can be paid in accordance to living costs in specific countries

	Supplier Ethical Standards	Factory Lists	Retailer Code of Conduct	Audit Process Details	Minimum Wage	Freedom of Association	Supply Chain Transparency
Company A	X	X	X	X	X	X	X

Fig. 3 Online CSR communication study—Company A results

- **Freedom of Association**—Giving workers the right to come together with others to collectively pursue common interests
- **Supply Chain Transparency**—Making all stages of the supply chain explicit and public.

The results of this online study and the company compliance with each criteria factor will be discussed by individual retailer.

Company A

When initially visiting the merchandise homepage of the Company A website, it appeared as a predominantly retail portal. However, a link to further details about the company including the CSR commitments and strategy can be found at the bottom of the homepage. This link redirects the user to the CSR micro-site, which is in keeping with the companies branding and aesthetically pleasing to the reader. The use of animations and striking photography makes the website engaging and informative. The delivery of information regarding the companies CSR strategy is delivered at a very readable and informal manner, allowing the user to interact with the site and the information provided on a number of levels.

Throughout the site, specific commitments are explained and where necessary related directly to a particular product area. However, due to the user being directed away from the merchandise site, the product and CSR values become very separated. Through the presentation of the ethical and sustainable issues and the products available to purchase being two separate websites, this may indicate to the user that there is no direct relationship between the CSR values being explained and the products that the company retails.

In terms of the criteria to which the website is being analysed under, Company A appears very poor in their communication delivery. Whilst the CSR micro site delivers a lot of information regarding the CSR commitments and action the company as a whole is undertaking, the specific details that are required during this study are not made available. Across all seven criteria points Company A was give a red cross-indicating that no information of that nature was provided to their customers online (Fig. 3).

	Supplier Ethical Standards	Factory Lists	Retailer Code of Conduct	Audit Process Details	Minimum Wage	Freedom of Association	Supply Chain Transparency
Company B	✓	X	O	X	O	✓	O

Fig. 4 Online CSR communication study—Company B results

Company B

The main website of Company B illustrated the brand identity and nature of the merchandise superbly, with all links and subsequent pages reflecting the nature of the business. The website throughout comes across as very product focused, speaking about specific manufacturing techniques where necessary. The CSR strategy of the company features as one of the main headings on the homepage of the merchandise website, again linking the product directly to the ethics and sustainability information provided.

Specific details such as the criteria factors featured in a series of documents that can be downloaded. These are presented predominantly as PDF Word files, however, the brand imagery and logos are again evident. Whilst this method communicates some of the CSR information, the presentation format does not appear to do the work being carried out in terms of ethics justice. A more visual communication format may have been more appropriate in order to explain their actions in detail.

From the seven criteria factors, Company B provides information for all but two; factory lists and audit process details, details for which could not be found. Within the company code of conduct, supplier ethical standards are detailed, meaning that a green tick was given. With retailer code of conduct, minimum wage and supply chain transparency, certain levels of information were given, but no specific details were provided. These three criteria were therefore classified as to some extent. The final criterion; freedom of association, was featured in detail on the website meaning that a green tick was again given (Fig. 4).

Company C

Company C was the overall best rated company website for the communication of ethical and sustainable commitments. The aesthetics of the site are not an attractive as with Company B, with the branding of the retailer not being as iconic and recognisable. Again as with Company A, the consumer is directed away form the main merchandise website to a more business focused and corporate micro-site. This could again imply that that product and the CSR strategy are two separate concerns for the business, indicating a lack of connectivity.

	Supplier Ethical Standards	Factory Lists	Retailer Code of Conduct	Audit Process Details	Minimum Wage	Freedom of Association	Supply Chain Transparency
Company C	✓	X	✓	✓	○	✓	X

Fig. 5 Online CSR communication study—Company C results

	Supplier Ethical Standards	Factory Lists	Retailer Code of Conduct	Audit Process Details	Minimum Wage	Freedom of Association	Supply Chain Transparency
Company D	✓	X	✓	✓	X	✓	○

Fig. 6 Online CSR communication study—Company D results

The details provided for the criteria factors predominantly appeared in Word document format that could be downloaded if the reader wished. This communication format, despite provided some of the details required, appeared quite dull and unappealing to a generic customer, which may discourage them from gaining knowledge of details such as minimum wage standards as implemented by the company. Other details provided were through the annual sustainability reports, which were also both readable and downloadable for the customer.

When referring to the information search criteria, supplier ethical standards, retailer code of conduct, audit process details and freedom of association were all given a green tick as full details regarding these were provided. The number and type of factories used by Company C were detailed, yet no specific factory details were provided. This criterion was therefore given an orange circle, indicating that this information had been given to some extent. Similarly with the minimum wage category, the legal minimum wage was mentioned, however, the differing country specific wages were not detailed. The categories where no information was provided were supply chain transparency and factory lists (Fig. 5).

Company D

As with Company A and C, Company D directs the user away form the merchandise website in order to provide more details regarding their ethical and sustainable actions. The user is directed to a corporate site that is the umbrella site for all the brands held under Company D's corporate group. All brands are covered with the group code of conduct where no brand specific or product specific details are given. The corporate site over all is heavily focused on the business behind the brand and is very unappealing visually. Also the brand identity of the group is different from the specific brand (Company D) being focused on during this study. This again shows to the user a lack of connectivity between the merchandise being sold and the CSR strategy. The tone and language used also indicates that the target audience is different from that of the product being sold.

The criteria details being analysed were predominantly seen in a series of code of conduct documentation, presented as a PDF Word document. This was very text heavy, with no engaging imagery or brand identity portrayed. The documents also feature large quantities of jargon, which the reader would more than likely not understand. It appears that these documents are for industry professionals or individuals who have a substantial level of prerequisite knowledge regarding ethics and sustainability in the apparel industry.

Supplier ethical standards, retailer code of conduct, audit process details and freedom of association were all detailed in the word PDF word documents provided and therefore were given a compliant classification in the study. Counter wise no details were provided the factory lists and minimum wage categories. As with Company B, generic information was provided regarding the supply chain, however many process details were omitted (Fig. 6).

Company E

As with four of the five retailers studied, the user of the Company E website is navigated away from the merchandise website in order to read information about the CSR business strategy. The micro-site reflects well the brand image and features lots of imagery of the garment collections currently being sold. This link between product and CSR has not been seen in many of the websites, with Company E being the most successful at illustrating this link to their customers.

The information is provided using a number of different channels, such as text, imagery and video, making the details easy to access and interesting. A good balance of statistics and narrative also makes the site interesting yet informative.

	Supplier Ethical Standards	Factory Lists	Retailer Code of Conduct	Audit Process Details	Minimum Wage	Freedom of Association	Supply Chain Transparency
Company E	✓	X	✓	✓	X	✓	O

Fig. 7 Online CSR communication study—Company E results

Whilst a lot of information is made available to the user, navigation through the site is fairly complex with headings not being clear on the information being given.

Supplier ethical standards, retailer code of conduct, audit process details and freedom of association were all detailed in either the code of conduct document or annual sustainability report. However, as with company D there were no details provided for factory lists and minimum wage. For the supply chain transparency, however, a summary of roles within the process were given but with no further or specific information (Fig. 7).

Results—In-Store Analysis

The collection of this data was carried out from the point of view of a customer. Each store was individually visited with the criteria factors in mind, observing if the retailer used each method as a communication tool. Once this was established, the type of information provided was observed and the details were recorded.

The communication methods focused on for the purposes of the study were as follows:

- **Swing Tags**—Labels attached to the outside of a garment, usually featuring price and brand name
- **Care Labels**—Fabric labels which are attached in the side seams of a garment
- **Carrier Bags**—Any plastic or paper bags which are given to customers to carry their bought goods
- **Leaflets/magazines**—Any paper based publications made available to customers in-store
- **Window Displays**—Any posters, window vinyl or visual merchandising able to be seen from outside the store
- **Posters/advertisements**—Any interior decorations or posters able to be seen in-store
- **Initiatives**—Events or concessional ranges featured in-store to promote the company's CSR strategy.

The results of the in-store investigation and the methods used by the five retailers will be discussed individually.

	Swing Tags	Care Labels	Carrier Bags	Leaflets / Magazines	Window Displays	Posters / Advertisements	Initiatives
Company A	O	✓	✓	✓	X	✓	✓

Fig. 8 In-store CSR communication study—Company A results

Company A

When visiting the store of Company A, the presence of their CSR branding was obvious from the outset. In-store visuals, posters and advertisements could be seen on the walls and at the point of sale. The launch of a recent initiative could be seen in several places around the womenswear merchandising department, which was also featured in the monthly in-store magazine provided to customers.

The overall branding of the CSR commitments as mentioned were visible in many areas of the store, however, specific commitments or actions were less obvious. Within the womenswear department, these were not evident in terms of in-store visuals or advertisement, however, certain ethical commitments were shown on swing tags and garment care labels. On all garment care labels the message *wash at* 30° was given as an environmental message, reflecting the commitments made within their sustainability strategy. Also featured on care labels was a message regarding end of life disposal, reminding the customer to contribute their unwanted items of clothing to their in-store, recycling initiative. Swing tags were used as a method of communication only on garments featuring fair trade cotton, where the Fairtrade logo was featured to inform customers of the ethical implications of the select pieces. This directly reflected the company's social commitments. Where specific commitments were made explicit or directly quoted from the annual report was within the food department, however this approach did not feature across departments in-store.

Other methods of in-store communication evidenced within Company A were the carrier bags given to customers when making a purchase. Featured on the bags were the CSR commitment branding and the associated website address. Also featured was a statement informing the customer that the plastic bag was made from 100 % recycled plastic and to reuse the bag provided to protect the environment. The only criterion within the communication study that Company A did not utilise in-store was window displays as part of CSR (Fig. 8).

Company B

When visiting Company B, the branding of the company and the aesthetic philosophy was very obvious in the store decoration and colour scheme used

	Swing Tags	Care Labels	Carrier Bags	Leaflets / Magazines	Window Displays	Posters / Advertisements	Initiatives
Company B	✓	X	✓	✓	X	✓	✓

Fig. 9 In-store CSR communication study—Company B results

throughout. Evidence of the CSR commitments could be seen in several areas of the store, with the branding of these commitments making this easier for the customer to identify.

What Company B featured within their in-store communication, which none of the other companies surveyed did, was the inclusion of CSR branded swing tags on every garment. These were separate swing tags to those featuring the price, bar code and product specific information and included no other information apart from the explanation of the ethical and sustainable philosophy of the company. These tags were of a high quality and due to their presence on every garment in-store, reflected the company's emphasis and importance of the CSR commitments made. The branded CSR logo was also featured on every carrier bag given to customers. A lot of the merchandise made available by Company B, is of a heavily handcrafted, embellished nature. How the manufacturing supply chain manages the social responsibility surrounding this handwork is also explained in swing tag format on the relevant garments. This is again a feature of CSR communication that is not used by the remaining four retailers surveyed.

A further feature that is unique to Company B is the use of small informative signs positioned next to certain products to inform customers of a specific ethical attribute. For example, garments that included fair trade materials were positioned next to a sign giving customers further information of the nature of the fair trade materials used. Another example is merchandise that is associated with a charity partnership, explaining that all profits from specific products are donated directly to charity organisations working in partnerships with the brand. These small signs interact with consumers on a level where the individuals can choose to read as little or as much information as they want, leaving the customer with the choice. Within the criteria of the communication study, this partnership provided information for the initiatives section (Fig. 9).

Company C

When initially carrying out the study in the store of Company C, no CSR communication methods were immediately obvious. On further investigation, a branded ethical trade sign was positioned behind the till points. This sign, however, provided no further information regarding how this was being carried out or indeed what it entailed. Whilst this offered customers a simple message, no details were given directing the customer to a source of further information.

Another method of communication used in-store was the feature of a CSR commitment on the care labels of selected garments. As seen with Company A, this instructed customers to *wash at 30°, if the garment is not dirty*. This commitment was again in alignment with the environmental strategy made in the company's annual report.

The final method used by Company C to communicate their CSR message to customers was through their catalogue available for customers to take home. This included several pages at the back of the publication explaining the commitments and actions the company is presently taking towards social and environmental responsibility (Fig. 10).

	Swing Tags	Care Labels	Carrier Bags	Leaflets / Magazines	Window Displays	Posters / Advertisements	Initiatives
Company C	X	✓	X	✓	X	✓	X

Fig. 10 In-store CSR communication study—Company C results

Company D

Upon visiting Company D's store, (as with Company C), no communication methods were immediate obviously. Working to the criteria of the study, only one of the seven observed elements was found to be present. The care labels featured in the garments displayed a similar message to that found with Company A and C; *save energy, wash at* 30°.

The lack of communication evidence found indicates that the brand are either doing very little to communicate their CSR message to their customers, or indeed they are doing relatively little towards a more social or environmentally responsible future. With the exception of care labels, the findings show that they are not communicating any social action to their customers through any communication channel found during the data collection process (Fig. 11).

Company E

Company E were unique in the way they communicate their CSR message to their consumers due to their approach being to highlight their actions through a specific collection. This range had a different visual branding identity to the regular collections found in-store, portraying a more in-touch with nature image. This collection featured swing tags explaining the environmental action made through the production of the range. When looking to the remaining collections found in-store, no swing tags indicating CSR commitments or actions were used. This strategy of communication relied heavily on the specific collection to inform the consumer of

	Swing Tags	Care Labels	Carrier Bags	Leaflets / Magazines	Window Displays	Posters / Advertisements	Initiatives
Company D	X	✓	○	X	X	X	X

Fig. 11 In-store CSR communication study—Company D results

the company's CSR activities, where using a multi-channel approach may have improved this.

The final method of communication found to be undertaken by Company E was a coding process found within their catalogue publications. Next to garments that directly reflected one or more of the ethical or sustainable commitments made by the retailer, a green dot was marked. When referring to the details in the back of the catalogue, this informed customers that the garment was in some way compliant with their CSR strategy. The details of the method of compliance or the differentiation between social or environmental attributes were not indicated. This illustrates that within the leaflets/magazines criterion of the communication, Company E complied with some extent. This was due to the commitment being visualised to the customer, however, no further detail being given (Fig. 12).

4 The Creation of Connectivity, Understanding and Empathy

4.1 The Current Situation

As discussed throughout the chapter, there remains an issue with consumer knowledge and awareness of the garment supply chain. Despite this being a long and complex process, consumers remain in the dark about where and how the garments they purchase are made. Not only do they not know about the complexities of the supply chain, but they also remain unaware of the social and environmental complexities which can occur. This lack of understanding of the supply chain creates disconnect between the consumers purchasing the fashion products and the suppliers who make the goods.

When considering the fashion supply chain, three main stakeholders can be identified; consumers, retailers and manufacturers. The inclusion of these parties sees the entirety of the supply chain accounted for, from the production of the goods, to the marketing and retail and finally to the consumer who chooses to purchase specific items. There is only one stakeholder, however, who has a relationship with both others and that is the retailers. Retailers have a relationship with the manufacturers as the role of the client who ultimately employs the manufacturer

	Swing Tags	Care Labels	Carrier Bags	Leaflets / Magazines	Window Displays	Posters / Advertisements	Initiatives
Company E	✓	X	X	○	X	X	✓

Fig. 12 In-store CSR communication study—Company E results

to produce their goods. As the client the retailer has a certain amount of power and leverage to demand certain conditions as part of their business agreement. These factors identified in the supply chain could include the factory producing the goods, methods of transportation and the date for delivery ready for retail sales. Likewise with the consumer, the retailer has a relationship during the purchasing process where the company will implement marketing strategies to encourage sales. This two-way relationship leaves the retailer in a powerful position ultimately having influence over both the manufacturing supply chain and the provision of fashion to consumers during the purchasing process. This power is extended through further collaboration with other retailers and brands enabling both companies leverage and further power in the industry. Whilst being currently underused, it is this powerful position that retailers can use to help move the fashion industry to a more socially responsible and sustainable future. The level of influence they have could not only make changes in their relationship with manufacturers but through the implementation of marketing strategies, thus being able influence consumers to also purchase more responsibly.

Despite the retailer's powerful position having positive connotations for the development of the industry in the future, in its current state, retailers are acting as a barrier between other key stakeholders having a relationship. The retailer acting as the middleman communicates and influences their business with both manufacturers and consumers, whilst manufacturers and consumers maintain no contact. This barrier is a key contributing factor to the popularity of the ethical fashion market as consumers have no appreciation and understanding of the origin of their fashion purchases made (Fig. 13). As previously discussed, very few consumers are knowledgeable to a level where an informed socially responsible purchasing decision can be made. With little to no appreciation for the conditions or culture in which their garments are made, consumers have no connectivity with the people producing their clothing and as a result little empathy with their social situation (James 2015).

The current CSR communication conducted by high street fashion retailers comes with several issues attached. The first issue is the lack of communication itself. As evidenced during the case study example, the breadth of level of communication varies dramatically, with some retailers sharing a lot of information regarding their CSR strategy and others shying away from the public exposure of their practices. As previously discussed there remains an issue in the industry of the consumer not having enough knowledge and awareness regarding social and environmental responsibility. With the retailer in a prominent position, CSR communication could aid in not only increasing this knowledge but also raising levels of engagement in such practices. There is a clear link between increased CSR communication and the informed consumer; this shall be discussed in more detail in the following section of the chapter.

Another key issue identified in the current use of CSR communication is the methods practiced by companies to inform and engage their customers. As identified during the case study example, the in-store environment is a very underused opportunity for companies to communicate to consumers. Very few of the

Fig. 13 The relationship of the key stakeholders in the fashion supply chain (James 2015)

companies studied took advantage of this opportunity, yet this is where many consumers spend the majority of their time browsing, potentially being open to engaging with social marketing messages. Methods of online communication, often misses the mark with engaging the reader. As seen with company D, communication comes in the form of a jargon and text heavy document aimed predominantly at stakeholders and not consumers. This approach to CSR communication is detrimental to its success and can put consumers off interacting with such topics in the future. The level in which a company attempts to engage a consumer is vital, with the company needing to understand the needs and wants of consumers when it comes to social and environmental responsibility.

4.1.1 Recommendations for the Future

A change in this relationship between consumers and manufacturers could encourage a more positive level of knowledge and understanding from the fashion customer and in turn influence their behaviour to that of more socially responsible choices. Adopting this approach could benefit all three parties involved,

- **Suppliers**—As mentioned earlier in the chapter, recent years have seen many social disasters such as the collapse of the Rana Plaza garment factory happen as a consequence of pressure being put on the supply chain as a result of the growth of the value sector and the fast fashion market. A change in consumer purchasing to more socially responsible product would help towards the prevention of such events reoccurring and aid in the development of better working conditions and fair living wages for garment workers.

- **Consumers**—Currently consumers are often presented with a choice, where they have to make trade off decisions between products which are ethical or produced under good social conditions, and those which are potentially cheaper or more on trend but where details of the supply chain are often not detailed. A conscious change made by retailers would eliminate this series of trade offs that often occur and consumers would be able to put their trust in a retailer to act responsibly.
- **Retailers**—As discussed throughout this investigation, high street retailers have been adapting business practices in order to be conducting their business in both a more socially and environmentally friendly way. Each of the five retailers involved within this research have CSR strategies currently being implemented and actively push these forward for further development and future activity. By improving the current relationship between their suppliers and consumers, retailers will have more demand for ethical product and by opposing the segregation of such products, can begin to implement this throughout their business practice.

Some brand and retailers have started this journey of involving the customer at every stage of the supply chain. Narrative has played a big role in telling the story behind a product, how it is manufactured and often the people involved in bringing that product to market. As previously discussed, companies such as Made-by with their Mode Tracker are beginning to connect consumers with the supply chain, with a personalised approach being more effective as each supply chain varies in details and complexity. Furthermore, recently launched social initiative Durated (www. durated.com), described as *an informed shopping experience for high quality design, designed by stories*. This website aims to provide shoppers with the narrative of the supply and manufacture of goods prior to purchase. The way this is presented is a real unique selling point, further reinforcing that customers want to know the story behind products they purchase.

In order for CSR communication to be effective in its objective to inform the consumer, several key approaches need to be considered. These recommendations are as follows:

- A direct connection to be made between product and CSR message.
  From the online and in-store communication study carried out, it was seen that when retailers communicate CSR messages to their customers, they make very little connection between the message being shared and the products being sold. This if often due to the method of communication being used or that users are diverted away from the merchandise website to a micro, more business-orientated site. Retailers could not only increase their level of communication here but also begin to engage the customer further through the product aesthetics. This approach would better inform consumers of social issues affecting the garment supply chain, allowing them to make a more informed decision. As a relatively simple change to make through a series of small changes such as CSR information being better position on the product

website, this is a short-term change that could be implemented quickly for maximum benefit.

- CSR messages to be communicated using product aesthetics as a vehicle.
  Throughout this research the idea that consumers purchase predominantly for aesthetics has been a reoccurring theme. This has also been highlighted in the problems found, where one of the factors inhibiting further ethical purchasing is a lack of trend-led or fashionable product available. Whilst the segregation of ethical and non-ethical is not encouraged, the communication of ethical messages through aesthetics could be a short-term solution that could be implemented relatively quickly. The Eco-Conscious Collect at H&M would be a good example of this, where the sustainable message is brought to the attention of customers through a limited range collection. Not only would this approach increase the retailer's communication of CSR messages, using specific ranges or collections could aid in growing awareness and consumer knowledge until more long-term and innovative solutions can be put into place.

- The use of innovative methods and approaches to better communicate ethical or social issues in fashion.
  Evidence from the primary collection process has shown that ethics needs to be disseminated to a wider audience with care in order to avoid green washing or to appear to be using ethics as a marketing tool. Whilst swing tags and communication through aesthetics can provide relatively short-term more effective methods, long term, retailers need to think of more innovative approaches to really engage consumers and push the industry forward. With this innovative approach, retailers could begin to communicate social issues in alignment with more environmental focused problems, moving away from sustainability concerns being more readily publicised. This approach could also begin to differentiate between the key differences between environmental and social issues, providing consumers with further knowledge to address their confusion.

- CSR messages to be simple and communicated in an easy to engage manner.
  Again this is an adjustment of current communication methods that is relatively simple can be implemented in the short term. Consumers and retailers alike believe that any CSR messages or activity shared should be simple, allowing the consumer to engage at a level that is best suited to them. It is simple, to the point pieces of information that the consumer will retain for future use. This action aims to address the consumer confusion regarding the term ethical, and again begin to distinguish ethical issues from environmental ones.

- Retailers to use their position to create relationships between consumers and suppliers.
  As previously discussed, retailers are in a unique and powerful position in the industry, with the ability to influence both consumers in the purchasing process and suppliers through the supply chain. However they also act as a barrier between consumers and suppliers, resulting in the two parties having no relationship or understanding of the other. This is potentially having a negative effect on ethical fashion purchasing, as consumers have no connectivity or

empathy with the people who produce the clothes they purchase. This research believes that retailers through further communication of CSR messages, using some of the innovative methods previously discussed, could help in the development of this relationship. This interrelationship development is a long-term commitment that would need the full cooperation and engagement of both retailers and suppliers. With this increased knowledge of suppliers, it is hoped that consumers would begin to purchase with their true moral values and apply their conscience when making purchasing decisions.

- The development of relationships with suppliers in order to create value in ethical attributes.

  This action reflects the one discussed previously, where a case was made for the development of a relationship between consumer and supplier. This action however lays emphasis on how this relationship could bring value to ethical attributes in garments, and the consumer's appreciation of this. From the research found within this investigation, it can be seen that consumers purchase predominantly for aesthetics and whilst ethics is a desirable attribute, it is not currently seen as essential. This indicates a lack of consumer value in ethical attributes, however through increased awareness and insights into the supply chain, it is thought that this value in socially responsible manufacturing would be developed. As this stage relies on the development and implementation of the relationships between consumers and suppliers, it can be seen as a long-term action that will rely on other incremental steps discussed before it can be successful.

When taking these recommendations into consideration during the communication process, companies can begin to really engage consumers in their values, whilst further informing them of social and environmental responsibility in the fashion industry. Through this effective communication, companies can begin to aid in breaking down the knowledge barrier leading to more informed fashion consumers.

5 Conclusion

5.1 Summary

In Sect. 1.1.1 it was stated that it is believed that the very act of purchasing is a consumer's expression of their own individual judgment and values, however, this can work both ways, with consumers also choosing not to shop somewhere due to disagreeing with a companies behaviour (Smith 1995). Throughout the chapter this has remained the key pivot point to focus on, as purchasing attitude is what still motivates profit driven retailers. If the consumer became more knowledgeable, retailers would have to think a lot more about whether to integrate CSR.

Throughout the research the topic of consumer supplier relationship is mentioned consistently, explored in some depth however this relationship between supplier and consumer remains difficult as they are often at other ends of the globe and working on different schedules and timescales.

In contrast marketing an ethical product is easy when the supply chain is transparent, however marketing the impression of having an ethical supply chain and being ethical on the surface is also all too easy. A more informed consumer would therefore benefit from being able to make better informed decisions as to whether a retailer and its product met the responsible CSR criteria.

Previously mentioned in Sect. 2.1.1 retailers also control how much information they reveal to the public about their supply chains and the methods utilised in the manufacture of their products. This transparency, or lack of it in many cases, hides the amount of information revealed about the suppliers to produce the garments. The fact that it varies greatly from company to company is in fact part of the problem as it means there are no strict guidelines to control behaviour.

The research analysis has highlighted in the company case study on CSR communication that the gap between what a company says they are doing in terms of CSR actions and what they are actually doing has been called *corporate hypocrisy*, with perceived hypocrisy said to be having a negative effect on consumer attitudes towards retailers (Wagner et al. 2009).

As discussed throughout the chapter very few consumers are knowledgeable to a level where an informed socially responsible purchasing decision can be made. With little to no appreciation for the conditions or culture in which their garments are produced, consumers have no connectivity with the people producing their clothing and as a result little empathy with their social situation (James 2015). The lack of communication evidence found indicates that the brands are either doing very little to communicate their CSR message to their customers, or indeed they are doing relatively little towards a more social or environmentally responsible future. The findings show that they are not communicating any social action to their customers through any communication channel found during the data collection process.

As in many aspects of life there are leaders and followers, in this case both H&M and Patagonia in Sect. 2.1.1 have in different ways taken a stance to lead for change but there are gaps, mainly in the lack of integration of CSR within day-to-day company strategy and planning and production processes.

It was only really the bar coding method as demonstrated by companies such as Made-by that has shown any signs of making progressive change of attitude. As technology advances this should in fact come into more use through default. This development is believed to be the key in creating understanding and empathy in order to influence socially responsible fashion purchasing.

References

Andreasen, A. (1995). *Marketing social change. Changing behavior to promote health, social development and the environment.* San Francisco: Jossey-Bass.

Aras, G., & Crowther, D. (2009). Corporate sustainability reporting: A study in disingenuity? *Journal of Business Ethics, 87,* 279–288.

Burchell, J. (2008). *The corporate social responsibility reader: Context & perspectives.* Routledge.

Cerin, P. (2002). Communication in corporate environmental reports. *Corporate Social Responsibility and Environmental Management, 9,* 46–65.

Compare My Spend. (2012). Clothing and footwear. Available at: http://www.comparemyspend.co.uk/clothing-footwear.php. Accessed: April 1, 2016.

Devinney, T. M. (2009). Is the socially responsible corporation a myth? The good, the bad, and the ugly of corporate social responsibility. *The Academy of Management Perspectives, 23,* 44–56.

Ethical Trading Initiative. (2010). Making the invisible workforce visible: Tracking homeworkers in Monsoon accessorize' supply chain. Available at: http://www.ethicaltrade.org/sites/default/files/resources/Case%20study_Monsoon.pdf. Accessed: April 4, 2016.

Fox, K. (2005). *Watching the english.* London: Hodder & Stoughton.

Gentleman, A. (2007). Gap moves to recover from child labour scandal. Available at: http://www.nytimes.com/2007/11/15/business/worldbusiness/15iht-gap.1.8349422.html?_r=0. Accessed: March 11, 2016.

Harney, A. (2008). *The China price—The true cost of chinese competitive advantage.* New York: The Penguin Press.

Hartley-Brewer, J. (2015). Be honest: you don't care if your pretty dress was made by child slaves. Available at: http://www.telegraph.co.uk/news/worldnews/asia/vietnam/11770335/Be-honest-you-dont-care-if-your-pretty-dress-was-made-by-child-slaves.html. Accessed: March 12, 2016.

Hawkes, S. (2013). People thought Rana Plaza would be a blow to Primark. Today's profit figures say otherwise. *The Telegraph, 5.* Available at: http://blogs.telegraph.co.uk/news/stevehawkes/100244468/people-thought-rana-plaza-would-be-a-blow-to-primark-todays-profit-figures-say-otherwise/. Accessed: January 30, 2013.

Holme, R., & Watts, P. (1999). *Corporate social responsibility.* Geneva: World Business Council for Sustainable Development.

Ihlen, Ø., Bartlett, J. L., & May, S. (2011). Corporate social responsibility and communication. In Ø. Ihlen, J. L. Bartlett, & S. May (Eds.), *The Handbook of Communication and Corporate Social Responsibility* (pp. 1–22). New Jersey: Wiley-Blackwell.

James, A. M. (2015). *Influencing ethical fashion consumer behaviour—A study of UK fashion retailers.* Unpublished Ph.D. thesis. University of Northumbria at Newcastle.

Kirk, A. (2015). How much does the average household spend each week. Available at: http://www.telegraph.co.uk/property/uk/how-much-does-the-average-household-spend-each-week/. Accessed: April 6, 2016.

Klein, N. (2014). Naomi Klein: The hypocrisy behind the big business climate change battle. Available at: http://www.theguardian.com/environment/2014/sep/13/greenwashing-sticky-business-naomi-klein. Accessed: March 16, 2016.

Kotler, P., & Lee, N. (2005). *Corporate social responsability. Doing the most good for your company and your cause.* New Jersey: John Wiley & Sons.

Lindgreen, A., & Swaen, V. (2010). Corporate social responsibility. *International Journal of Management Reviews, 12,* 1–7.

Made-By. (2016). Four brands take pioneering first step. Available at: www.made-by.org/news/press-release/major-step-forward-for-fashion-industrys-sustainability-drive/. Accessed: April 1, 2016.

Miller, C. (2009). Sustainable marketing and the green consumer. In E. Parsons & P. Maclaren (Eds.), *Contemporary issues in marketing and consumer behaviour* (pp. 141–160). Oxford: Butterworth—Heinemann.

Niinimaki, K. (2010). Eco-clothing, consumer identity and ideology. *Sustainable Development,* *18,* 150–162.

Patagonia. (2011). Don't buy this jacket. Available at: http://www.patagonia.com/email/11/112811.html. Accessed: March 22, 2016.

Smith, N. C. (1995). Marketing strategies for the ethics era. *Sloan Management Review, 36,* 85–98

Smith, N. C., & Ward, H. (2007). Corporate social responsibility at a crossroads? *Business Strategy Review, 18,* 16–21.

Strom, S. (1996). A sweetheart becomes suspect; looking behind those Kathie Lee labels. Available at: http://www.nytimes.com/1996/06/27/business/a-sweetheart-becomes-suspect-looking-behind-those-kathie-lee-labels.html?pagewanted=all&src=pm. Accessed: March 22, 2016.

Vogel, D. J. (2005). Is there a market for virtue? The business case for corporate social responsibility. *California Management Review, 47,* 77–92.

Waddock, S., & Googins, B. K., (2011). The paradoxes of communicating corporate social responsibility. *The handbook of communication and corporate social responsibility* (pp. 23–43).

Wagner, T., Lutz, R. J., & Weitz, B. A. (2009). Corporate hypocrisy: Overcoming the threat of inconsistent corporate social responsibility perceptions. *Journal of Marketing, 73,* 77–91.

Wales, A., Gorman, M., & Hope, D. (2010). *Big business, big responsibilities: From villains to visionaries: How companies are tackling the world's greatest challenges.* Palgrave Macmillan.

Zadek, S., & Chapman, J. (1998). *Revealing the emperor's clothes: How does social responsibility count?* London: New Economics Foundation.

Printed in the United States
By Bookmasters